PRINCETON AERONAUTICAL PAPERBACKS

1. LIQUID PROPELLANT ROCKETS
David Altman, James M. Carter, S. S. Penner, Martin Summerfield.
High Temperature Equilibrium, Expansion Processes, Combustion of Liquid Propellants, The Liquid Propellant Rocket Engine.
196 pages. $2.95

2. SOLID PROPELLANT ROCKETS
Clayton Huggett, C. E. Bartley and Mark M. Mills.
Combustion of Solid Propellants, Solid Propellant Rockets.
176 pages. $2.45

3. GASDYNAMIC DISCONTINUITIES
Wallace D. Hayes. 76 pages. $1.45

4. SMALL PERTURBATION THEORY
W. R. Sears. 72 pages. $1.45

5. HIGHER APPROXIMATIONS IN AERODYNAMIC THEORY. M. J. Lighthill.
156 pages. $1.95

6. HIGH SPEED WING THEORY
Robert T. Jones and Doris Cohen.
248 pages. $2.95

7. FUNDAMENTAL PHYSICS OF GASES
Karl F. Herzfeld, Virginia Griffing, Joseph O. Hirschfelder, C. F. Curtiss, R. B. Bird and Ellen L. Spotz.
149 pages. $1.95

8. FLOW OF RAREFIED GASES
Samuel A. Schaaf and Paul L. Chambré.
63 pages. $1.45

9. TURBULENT FLOW
Galen B. Schubauer and C. M. Tchen.
131 pages. $1.45

10. STATISTICAL THEORIES OF TURBULENCE
C. C. Lin.
68 pages. $1.45

PRINCETON UNIVERSITY PRESS · PRINCETON, N.J.

NUMBER 7
PRINCETON AERONAUTICAL
PAPERBACKS
COLEMAN duP. DONALDSON, GENERAL EDITOR

FUNDAMENTAL PHYSICS OF GASES

BY K. F. HERZFELD, VIRGINIA GRIFFING,
J. O. HIRSCHFELDER, C. F. CURTISS, R. B. BIRD
AND ELLEN L. SPOTZ

PRINCETON, NEW JERSEY
PRINCETON UNIVERSITY PRESS
1961

L. C. CARD NO. 61-12663

PRINTED IN THE UNITED STATES OF AMERICA

HIGH SPEED AERODYNAMICS

AND JET PROPULSION

I. Thermodynamics and Physics of Matter. Editor: F. D. Rossini
II. Combustion Processes. Editors: B. Lewis, R. N. Pease, H. S. Taylor
III. Fundamentals of Gas Dynamics. Editor: H. W. Emmons
IV. Theory of Laminar Flows. Editor: F. K. Moore
V. Turbulent Flows and Heat Transfer. Editor: C. C. Lin
VI. General Theory of High Speed Aerodynamics. Editor: W. R. Sears
VII. Aerodynamic Components of Aircraft at High Speeds. Editors: A. F. Donovan, H. R. Lawrence
VIII. High Speed Problems of Aircraft and Experimental Methods. Editors: A. F. Donovan, H. R. Lawrence, F. Goddard, R. R. Gilruth
IX. Physical Measurements in Gas Dynamics and Combustion. Editors: R. W. Ladenburg, B. Lewis, R. N. Pease, H. S. Taylor
X. Aerodynamics of Turbines and Compressors. Editor: W. R. Hawthorne
XI. Design and Performance of Gas Turbine Power Plants. Editors: W. R. Hawthorne, W. T. Olson
XII. Jet Propulsion Engines. Editor: O. E. Lancaster

PRINCETON, NEW JERSEY
PRINCETON UNIVERSITY PRESS

PREFACE

The favorable response of many engineers and scientists throughout the world to those volumes of the Princeton Series on High Speed Aerodynamics and Jet Propulsion that have already been published has been most gratifying to those of us who have labored to accomplish its completion. As must happen in gathering together a large number of separate contributions from many authors, the general editor's task is brightened occasionally by the receipt of a particularly outstanding manuscript. The receipt of such a manuscript for inclusion in the Princeton Series was always an event which, while extremely gratifying to the editors in one respect, was nevertheless, in certain particular cases, a cause of some concern. In the case of some outstanding manuscripts, namely those which seemed to form a complete and self-sufficient entity within themselves, it seemed a shame to restrict their distribution by their inclusion in one of the large and hence expensive volumes of the Princeton Series.

In the last year or so, both Princeton University Press, as publishers of the Princeton Series, and I, as General Editor, have received many enquiries from persons engaged in research and from professors at some of our leading universities concerning the possibility of making available at paperback prices certain portions of the original series. Among those who actively campaigned for a wider distribution of certain portions of the Princeton Series, special mention should be made of Professor Irving Glassman of Princeton University, who made a number of helpful suggestions concerning those portions of the Series which might be of use to students were the material available at a lower price.

In answer to this demand for a wider distribution of certain portions of the Princeton Series, and because it was felt desirable to introduce the Series to a wider audience, the present Princeton Aeronautical Paperbacks series has been launched. This series will make available in small paper-backed volumes those portions of the larger Princeton Series which it is felt will be most useful to both students and research engineers. It should be pointed out that these paperbacks constitute but a very small part of the original series, the first seven published volumes of which have averaged more than 750 pages per volume.

For the sake of economy, these small books have been prepared by direct reproduction of the text from the original Princeton Series, and no attempt has been made to provide introductory material or to eliminate cross references to other portions of the original volumes. It is hoped that these editorial omissions will be more than offset by the utility and quality of the individual contributions themselves.

Coleman duP. Donaldson, General Editor

PUBLISHER'S NOTE: Other articles from later volumes of the clothbound series, *High Speed Aerodynamics and Jet Propulsion*, may be issued in similar paperback form upon completion of the original series.

CONTENTS

SECTION B

FUNDAMENTAL PHYSICS OF GASES

CHAPTER 1. QUANTUM MECHANICS AND APPLICATIONS TO MOLECULAR STRUCTURE

KARL F. HERZFELD
VIRGINIA GRIFFING

B,1. Historical Introduction. From the time of Newton, the three general laws of nature which bear his name were successful in describing and predicting the mechanical behavior of a wide class of phenomena ranging from the motion of the celestial bodies to that of a projectile on the surface of the earth. However, with the discovery of electrons and the study of the behavior of atoms and molecules these laws were found to be inadequate, and, in fact, sometimes gave results that were in direct contradiction to experimental facts. It was necessary to develop a new system of general principles which would describe the motions of small particles. The new system consists of a set of equations called quantum mechanics which are based on concepts which are profoundly different from those of classical mechanics; however, for the heavier bodies of everyday life the quantum mechanics gives the same results as Newtonian mechanics.

The first step in the development of the new system was due to the German physicist Max Planck [*1*] who proposed that energy of light emitted from a black body was emitted discontinuously as quanta. The magnitude of the quantum he proposed was equal to the frequency of the emitted light multiplied by a new constant of nature now called Planck's constant h. Einstein later extended Planck's assumption to a successful explanation of the photoelectric effect. In 1913, Niels Bohr used the idea of quantization and the value of Planck's constant to deduce quantitatively the spectrum of the hydrogen atom. Bohr's theory as extended by Sommerfeld gave a qualitative understanding of emission and absorption spectra of atoms and molecules and reconciled the Rutherford planetary atom with the stable existence of atoms in certain stationary states.[1] This new theory, called quantum theory of atoms, assumed that Newtonian mechanics gave the correct equations for the motion of electrons in stable circular or elliptical paths about a heavy positively charged nucleus. Only those orbits were stable for which the angular momentum was an integral multiple of $h/2\pi$. Qualitatively these ideas could be carried over to molecules, but the electrons in molecules no longer move in a field of spherical symmetry; furthermore the spectra of more complicated atoms and those involving more than one electron were only qualitatively explained. From the aesthetic point of view the quantum conditions were arbitrarily superposed on classical mechanics without any basis in fundamental principles. Finally, the Bohr theory assumed that an atom can absorb only a full quantum and absorb it instantaneously; however, according to the electromagnetic wave theory of light, the energy would have been spread evenly over a sphere in such a way that the intensity would be so small that it would take an atom a very long time to absorb a quantum. This difficulty disappears if one assumes that light travels in little packets, i.e. quanta, but then one must explain the observed diffraction and interference of light waves. Thus by 1923–24 an impasse had been reached.

The first break in these difficulties came in the dissertation of de Broglie in 1924. De Broglie associated a wavelength λ with every particle of mass m moving with a velocity v where $\lambda = h/mv$. This was verified experimentally by Davisson and Germer and G. P. Thomson who showed that beams of electrons could be diffracted, and later by Stern who showed that particles of atomic mass also exhibited wave properties [*2*].

When two sets of independent but reliable experiments give an apparent contradiction and there is no error in the logic, then the only solution to the problem is to reexamine the fundamental postulates of the theory upon which the contradictory evidence depends. Thus the fundamental postulates of Newtonian mechanics must be reexamined in the light of these new experiments.

Now, the fundamental postulate of classical mechanics which must be considered can be stated in terms of Eq. 2-4 or Eq. 2-7 given below which say: Given an isolated system of n particles having coordinates $q_1, q_2, \ldots, q_{3n}$ and conjugate momenta $p_1, p_2, \ldots, p_{3n}$, the integration of these equations and the experimental evaluation of the $6n$ constants of integration enable one to describe completely the system at any future time. The $6n$ constants of integration may be the values of the $3n$ coordinates q_i^0 and $3n$ momenta p_i^0 at some given time $t = t_0$.

Heisenberg was the first to point out that mechanical description of

[1] The knowledge of the Bohr theory is invaluable in associating intuitive ideas with the more abstract quantum mechanics. An excellent account of these developments from this point of view are given in [*1*].

the motion of the electron about the nucleus of an atom required the simultaneous knowledge of the position and momentum of the electron. However, any process that would enable one to determine exactly the position of the electron would cause the electron to separate from the atom, thus changing its momentum. For example, one might be able to observe the electron with a γ-ray microscope but the energy absorbed by the electron in the process ionizes the atom. Thus Heisenberg postulated that classical mechanics is not adequate for the description of atoms and electrons because it is impossible to determine precisely for a given particle both p_i and q_i simultaneously and that the best one can do is given by the following relation $\Delta p_i \Delta q_i \geqq h/2\pi$.

There are other pairs of conjugate variables, related by a similar indeterminacy equation: e.g. energy and time form such a pair. The uncertainty principle proposed by Heisenberg says that for absorption of strictly monochromatic light by an atom, which requires a determined energy $E = h\nu$, the moment of absorption completely eludes observation or more precisely

$$\Delta E \Delta t \geqq \frac{h}{2\pi}$$

This uncertainty for particles of the ordinary world is so small that it is less than the random disturbances and thus is important only for particles of small mass. Physically, the uncertainty principle says that one can determine the position of a particle precisely but then any previous knowledge of its corresponding momentum has been destroyed by the measurement of position; or, it is possible to determine the momentum precisely but any knowledge of the corresponding position has been destroyed in the process. If the determination of the position, for example, is made only between limits Δq, the measurement of q produces an unknown change of about Δp in the momentum.

The uncertainty relation explains the particle-wave duality by saying that it is impossible to follow a system and to describe it in sufficient detail to distinguish between the wave picture and the particle picture.

Thus the requirements of the new theory are that one must use the wave and particle pictures to describe the same system, while limiting the applicability of these pictures by the uncertainty principle. This leads to the natural introduction of statistical notions into the description of matter and it remains only to formulate a mathematical theory, the results of which can be correlated with experimental results by probability statements. The required mathematical theory was formulated almost simultaneously by Schrödinger and Heisenberg, using different mathematical tools.

Heisenberg believed that the Bohr theory failed because its fundamental quantities (orbits) cannot be measured. So he attempted to set up a system of atomic mechanics which introduced only observable quantities. That is, he assumed that the fundamental entities in the mechanics of atoms are the observable frequencies and their intensities. Using these quantities, Heisenberg, Born, and Jordan developed an entire science called matrix mechanics which has been a highly successful theory; however, it is complicated in mathematical formulation and is not here considered in any detail.

Schrödinger derived a "wave equation" by analogy between mechanics and physical optics which forms the basis of the new theory called wave mechanics. He showed shortly after the development of matrix mechanics and wave mechanics that the two were mathematically equivalent. Thus both are now called quantum mechanics. Because of its more familiar mathematical tools the Schrödinger method is used in this account.

B,2. Classical Mechanics. Newtonian mechanics can best be formulated in terms of Lagrange's equations which are convenient ways of writing Newton's second law and have the advantage that they hold in any sort of coordinate system, and not just in the Cartesian coordinate system. For simplicity consider a particle of mass m and Cartesian coordinates x, y, z, with the components of the velocity u, v, and w, respectively. Then consider that the potential energy is given by a function of the coordinates only, $V(x, y, z)$. Since

$$F_x = -\frac{\partial V}{\partial x}$$

the equations of motion, written in terms of the momenta, are given by

$$\frac{d}{dt}(mu) = -\frac{\partial V}{\partial x} \quad \text{where } u = \frac{dx}{dt} \tag{2-1}$$

as are similar equations for y and z. But the kinetic energy of the particle is given by T where

$$T = \frac{m}{2}(u^2 + v^2 + w^2)$$

thus

$$\frac{\partial T}{\partial u} = mu = p_x \tag{2-2}$$

where p_x is the linear momentum in the x direction and the equations of motion (2-1) can now be written

$$\frac{d}{dt}\left(\frac{\partial T}{\partial u}\right) + \frac{\partial V}{\partial x} = 0, \quad \frac{d}{dt}\left(\frac{\partial T}{\partial v}\right) + \frac{\partial V}{\partial y} = 0 \tag{2-3}$$

or defining the Lagrangian function $L = T - V$, Eq. 2-3 can be written as

$$\frac{d}{dt}\left(\frac{\partial L}{\partial u}\right) - \frac{\partial L}{\partial x} = 0$$

Now to generalize if a system is described by coordinates q_i and their time derivatives are given by $\dot{q}_i$ then the classical equations of motion become

$$\frac{d}{dt}\left(\frac{\partial L}{\partial \dot{q}_i}\right) - \frac{\partial L}{\partial q_i} = 0 \tag{2-4a}$$

where $L = T - V$, and T may now be a function of both q and $\dot{q}$ as occurs, for example, in the rotation of a particle. One may define a generalized momentum, belonging to q_i, by

$$\frac{\partial L}{\partial \dot{q}_i} = p_i \tag{2-4b}$$

Eq. 2-2 is a special case of Eq. 2-4b. Eq. 2-4a can be rewritten

$$\frac{d}{dt}\left(\frac{\partial L}{\partial \dot{q}_i}\right) = -\frac{\partial V}{\partial q_i} + \frac{\partial T}{\partial q_i} \tag{2-5}$$

or

$$\frac{dp_i}{dt} = -\frac{\partial V}{\partial q_i} + \frac{\partial T}{\partial q_i}$$

Since $\partial L/\partial \dot{q}_i$ defines a generalized momentum p_i, the right-hand side of Eq. 2-5 gives the sum of the component of the force derivable from a potential and the last term $\partial T/\partial q_i$ is a fictitious force which arises when Cartesian coordinates are not used and the kinetic energy is no longer a function of $\dot{q}_i$ only.

It is useful to extend this formulation of classical mechanics to an even more symmetric form. Assuming a conservative system then Eq. 2-5 can be replaced by the two relations

$$\frac{dp_i}{dt} = \frac{\partial L}{\partial q_i}, \qquad p_i = \frac{\partial L}{\partial \dot{q}_i} \tag{2-6}$$

and thus the n second order differential equations (2-5) are replaced by $2n$ first order differential equations in (2-6). But now instead of using the Lagrangian function L consider a function H, which in a conservative system is simply the total energy $T + V$ expressed as a function of the generalized coordinates q_i and momenta p_i; then it can be shown that Eq. 2-6 becomes

$$\frac{dq_i}{dt} = \frac{\partial H(p_i, q_i)}{\partial p_i}, \qquad \frac{dp_i}{dt} = -\frac{\partial H(p_i, q_i)}{\partial q_i} \tag{2-7}$$

These are called Hamilton's[2] canonical equations of motion; p_i and q_i are said to be canonically conjugate variables. The Hamiltonian function for a mass m and Cartesian coordinates x, y, z is

$$H = \frac{1}{2m}(p_x^2 + p_y^2 + p_z^2) + V(x, y, z)$$

The law of conservation of energy in this notation is written

$$\frac{dH}{dt} = \sum \left[\frac{\partial H}{\partial q_i} \dot{q}_i + \frac{\partial H}{\partial p_i} \dot{p}_i \right] = 0 \tag{2-8}$$

The equations of motion (2-7) can be rewritten in terms of a convenient function, Poisson's bracket, which is defined as follows: Given two functions of the variables $F_1(p_i, q_i)$ and $F_2(p_i, q_i)$, Poisson's bracket of F_1 and F_2 is written

$$[F_1, F_2] = \sum_{i=1}^{n} \left[\frac{\partial F_1}{\partial q_i} \frac{\partial F_2}{\partial p_i} - \frac{\partial F_1}{\partial p_i} \frac{\partial F_2}{\partial q_i} \right] \tag{2-9}$$

In this form the equations of motion become

$$\frac{dq_i}{dt} = [q_i, H], \qquad \frac{dp_i}{dt} = [p_i, H] \tag{2-10}$$

and one finds in classical mechanics the following important relations:

$$[q_i, q_j] = 0$$
$$[p_i, p_j] = 0$$
$$[q_i, p_j] = \delta_{ij}$$

B,3. Mathematical Concepts. The mathematical formalism which one uses in quantum mechanics is operator algebra. Since this is an unfamiliar branch of mathematics, a brief discussion of the fundamental rules and definitions are given here before giving the mathematical formulation of quantum mechanics. The proofs are not given but may be found in any standard book on quantum mechanics.

Operators. An operator is defined as a symbol for a rule which tells one how to make one function from another function, that is

$$\alpha\psi(q_1, q_2, \cdots, q_n) = \Phi(q_1, q_2, \cdots, q_n) \tag{3-1}$$

[2] The equivalence of Newton's laws, Lagrange's equations, and Hamilton's equations is proved in [*3*]. For Cartesian coordinates the equivalence is obvious. The Hamilton equations are not restricted to conservative systems and the definition of the Hamiltonian is more generally

$$H = \sum p_i \dot{q}_i - L,$$

Here, one says the operator α when applied to the function $\psi(q_1, q_2, \cdots, q_n)$ gives the function Φ. These operators may be of all types, simple multiplication operators, differential operators, etc. For example, if α_1 is the operator representing multiplication by q_2 from the left on $\psi = q_1^2q_2^2$ then $\Phi = q_2\psi$ or if α_1 is the differential operator $\partial/\partial q_i$ then

$$\frac{\partial}{\partial q_i}\psi(q_1, q_2) = 2q_1q_2^2$$

The sum of two operators α and β is defined by

$$(\alpha + \beta)\psi(q_1, q_2 \cdots q_n) = \alpha\psi + \beta\psi \tag{3-2}$$

The product of two operators α and β is defined by

$$\alpha\beta\psi \equiv \alpha[\beta\psi] \tag{3-3}$$

That is, the function ψ is first operated on by β and the resulting function is then operated on by α. The reverse of this process will in general give a different result. That is

$$\alpha\beta \neq \beta\alpha$$

e.g. if α is the multiplication operator q and $\beta = \partial/\partial q$, then

$$q\frac{\partial}{\partial q}\psi \neq \frac{\partial}{\partial q}(q\psi)$$

Thus one would say that, in general, operators do not commute and one calls the difference $[\alpha\beta - \beta\alpha]$ the commutator of the operators α and β. $\alpha^n\psi$, means α successively operates n times on the function ψ.

The operators used in quantum mechanics are linear operators. A linear operator is defined by the following rule

$$\alpha[\psi_1(q) + \psi_2(q)] = \alpha\psi_1(q) + \alpha\psi_2(q) \tag{3-4}$$

Only certain types of function $\psi(q)$ are of physical interest. These functions may be real or complex but they must be single-valued, finite, and continuous over the complete range of the variables, or at least quadratically integrable, that is

$$\int \psi^*(q)\psi(q)d\tau < \infty \quad \text{where } d\tau = dq_1, dq_2, \cdots, dq_n \tag{3-5}$$

where $\psi^*(q)$ is the complex conjugate of $\psi(q)$.

The only operators of interest in quantum mechanics have the property

$$\alpha\psi = \alpha^*\psi^* \tag{3-6}$$

the α being called a Hermitian operator. Linear combinations of Hermitian operators are also Hermitian; however, the products of two Hermitian operators are Hermitian only if the two operators commute.

Assume that $\psi(q)$ belongs to a class of well-behaved functions and that α is a linear Hermitian operator. If the result of the operation is the function back again multiplied by a constant, that is

$$\alpha\psi(q) = a\psi(q) \quad \text{where } a = \text{real constant} \tag{3-7}$$

then $\psi(q)$ is called an eigenfunction of the operator α and a is called an eigenvalue. In general there will be a set of eigenfunctions $\psi_i(q)$ which when operated on by the operator α will yield a set of eigenvalues a_i. Then the eigenfunctions associated with different eigenvalues are orthogonal. That is

$$\int \psi_i(q)\psi_j(q)dq = \delta_{ij}, \qquad \delta_{ij}\begin{cases} = 1, & i = j \\ = 0, & i \neq j \end{cases} \tag{3-8}$$

When two or more eigenfunctions yield the same eigenvalue, this is a degenerate function. However, it is always possible to choose linear combinations of degenerate eigenfunctions which are orthogonal, and the linear combination will also be an eigenfunction of the operator. Furthermore, if a complete set of orthogonal functions are eigenfunctions of two different operators, α and β, then α and β commute. Conversely, if α and β commute, there exists a set of orthogonal functions which are simultaneously eigenfunctions of both operators.

When the operator is a differential one with several variables, Eq. 3-7 gives a homogeneous linear partial differential equation which can often be split into total homogeneous differential equations with the proper choice of coordinate system for the particular boundary conditions. There the problem often becomes identical with the Sturm-Liouville boundary-value problem discussed below.

Development into orthogonal functions. For simplicity, consider a variable x which is defined for the interval a to b. Consider a function $\psi(x)$ of the type defined in Eq. 3-5 which is continuous, finite, and single-valued in the interval a to b. There is a theorem in analysis which says that it is always possible to find a series made up of given orthogonal functions which will approximate $\psi(x)$ as closely as one likes in the interval from $x = a$ to $x = b$. That is

$$\psi(x) = \sum_{i=1}^{n} C_i\phi_i(x) \tag{3-9}$$

where

$$\int_a^b \phi_j^*\phi_i dx = 0, \quad \text{i.e. orthogonal} \tag{3-10}$$

and

$$\int_a^b \phi_i^*\phi_i dx = 1, \quad \text{i.e. normalized} \tag{3-11}$$

It is necessary that the orthogonal functions ϕ_i satisfy the same boundary conditions as the function $\psi(x)$. The expansion is possible if the set is complete and the series converges. If the expansion is possible

$$C_n = \int_a^b \phi_n f(x)dx \tag{3-12}$$

and all values of C_n can be determined by successive integrations according to Eq. 3-12. The condition for normalization of the general function $\psi(x)$ then becomes

$$\sum_{i=1}^{n} C_i^* C_i = 1 \tag{3-13}$$

Examples of orthogonal functions ϕ_n are well known in classical physics where they are solutions of certain well-known linear homogeneous second order total differential equations of the form

$$\frac{d^2y}{dx^2} + [\lambda^2 - f(x)]y = 0 \tag{3-14}$$

where $f(x)$ is a known function continuous over the range $x = a$ to $x = b$, with boundary conditions of the form

$$\frac{dy}{dx}(a) - gy(a) = 0, \qquad \frac{dy}{dx}(b) - Gy(b) = 0 \tag{3-15}$$

where g and G are constants. Eq. 3-14 and Eq. 3-15 define a Sturm-Liouville boundary-value problem and well-behaved solutions to Eq. 3-14 called eigenfunctions exist for only certain values of the parameter λ called eigenvalues. One is usually able to find these solutions in terms of such well-known functions as Fourier series, Legendre polynomials, Laguerre polynomials, Hermite polynomials, etc.

Eq. 3-14 also has the property that if $F_1(x)$ and $F_2(x)$ are solutions of Eq. 3-14 for the same λ then any arbitrary linear combination

$$y = a_1F_1(x) + a_2F_2(x) \tag{3-16}$$

is also a solution.

B,4. Introductory Quantum Mechanics. The consequences of the uncertainty principle which must be reflected in the new formulation of mechanics can be stated briefly in terms of two principles.

1. Indeterminacy: Since an exact knowledge of initial conditions of a system is impossible, exact prediction of future behavior is no longer possible but must be replaced by methods of predicting the average behavior of a system or the probability of finding a system in a given state from the knowledge of the initial state allowed within the limits of the indeterminacy principle.

2. Complementariness: According to the indeterminacy principle, increased accuracy in the knowledge of one variable can be obtained only at the expense of knowledge of the conjugated variable, for example, the more accurately one determines q_j the more undetermined p_j becomes; therefore, the description of light from the wave point of view excludes the possibility of determining anything about its particle nature and vice versa.

Furthermore, the formulation of quantum mechanics is carried out according to the Bohr correspondence principle, which states that in the limiting conditions between microscopic and macroscopic systems the results of quantum mechanics and classical mechanics must converge to the same result.

The wave character of light, as shown by interference, is expressed mathematically ir such a way that the quantity to which the usual wave equation is applied is the amplitude, which may be positive or negative. Since it is this amplitude which appears in the linear wave equation and to which the principle of superposition applies, the amplitude of light from two slits, for example, is added and because the amplitude may be positive or negative there may be destructive interference. Only after the sum of amplitudes is formed is it squared to give the observable quantity (light intensity) which can never be negative. If, as in the case of noncoherent sources, the intensities are added, there is no interference and the wave character of light does not come into evidence.

Accordingly, a mathematical apparatus which is to express the wave-like properties of particles must apply its linear operators not directly to the density, which can never be negative, but to a quantity which behaves like an amplitude and may be positive or negative or even complex. From this quantity, the density must be formed by getting the square of the absolute magnitude; then the density, like the light intensity, will never be negative, and wavelike properties, like destructive interference, may still be exhibited.

Consider a system of n degrees of freedom which would be described classically by specifying the value of n coordinates $q_1, q_2, \ldots, q_n$ and their conjugate momenta $p_1, p_2, \ldots, p_n$. Now the state of the system is described by a well-behaved function $U(q_1, q_2, \ldots, q_n, t)$ which is called the wave function of the system and has the following properties: $U(q_1, \ldots, q_n, t)$ may be real or complex, may have positive, zero, or negative values, and is interpreted so that $U^*Ud\tau$ is the probability of finding the system in a volume element of configuration space $d\tau$. That is for a single particle described in Cartesian coordinates $U^*(x, y, z, t)U(x, y, z, t)dxdydz$ would be the probability of finding the particle in the volume element $dxdydz$. If one assumes the particle is limited to a finite space, then in order to get numerical results the wave functions must be normalized, i.e.

$$\int_\tau U^*Ud\tau = 1 \qquad (4\text{-}1)$$

This says with certainty only that the particle is some place in space. Thus the normalization of orthogonal functions which was done in classical physics for mathematical simplification takes on physical meaning in quantum mechanics.

In quantum mechanics physical quantity is no longer represented by an ordinary variable but to every observable quantity there is assigned a linear Hermitian operator.

A rule for determining the quantum mechanical operator is to write the corresponding classical function, $F(p_j, q_j)$, replace q_j by the multiplication operator q_j, and p_j by the differential operator

$$\frac{h}{2\pi i}\frac{\partial}{\partial q_j}$$

in such a way as to secure a linear Hermitian operator. When defined in this way q_j, p_j and any polynomial constructed from these operators will be linear Hermitian operators. However, products of p_j, q_j do not commute, that is,

$$p_k q_j - q_j p_k = \frac{h}{2\pi i}\,\delta_{kj}, \qquad \begin{array}{ll} \delta_{kj} = 1, & \text{for } k = j \\ \delta_{kj} = 0, & \text{for } k \neq j \end{array}$$

This follows from the definition of the operator p

$$\left[\frac{h}{2\pi i}\frac{\partial(q_j)}{\partial q_j} - q_j\frac{h}{2\pi i}\frac{\partial}{\partial q_j}\right]U = \frac{h}{2\pi i}U$$

This is the quantum analogue of the classical statement about Poisson brackets (see end of B,2). Functions which contain products of p_j and q_j must be symmetrized, e.g. into $\frac{1}{2}(pq + qp)$. In classical mechanics, p_j and q_j are ordinary numbers and consequently do commute. The uncertainty principle is included in the mathematical description by thus choosing pairs of operators corresponding to a coordinate and its conjugated momentum that do not commute.

Table *B,4* gives the operators which are used in this discussion.

Table B,4. Some common operators of quantum mechanics.

Observable quantity	*Classical operator*	*Quantum mechanics operators*
generalized space coordinate	q_i	q_i
generalized momentum coordinate	p_i	$\frac{h}{2\pi i}\frac{\partial}{\partial q_i}$
angular momentum in Cartesian coordinates	$M_z = m(x\dot{y} - y\dot{x})$	$\frac{h}{2\pi i}\left(x\frac{\partial}{\partial y} - y\frac{\partial}{\partial x}\right)$
Hamiltonian in Cartesian coordinates	$\frac{1}{2m}[p_x^2 + p_y^2 + p_z^2] + V(x, y, z)$	$-\frac{h}{8\pi^2 m}\nabla^2 + V(x, y, z)$
some arbitrary function F	$F(p^n, q^m, pq)$	$F\left[\left(\frac{h}{2\pi i}\frac{\partial}{\partial q}\right)^n, q^m, \frac{1}{2}(pq + qp)\right]$

Consider a system, the state of which is described by a wave function ψ_i. If this wave function ψ_i is an eigenfunction of the operator β, this means mathematically that the operation β applied to ψ_i makes out of ψ_i the expression $b_i\psi_i$, b_i being the eigenvalue.

One now makes the physical assumption that a measurement of the observable quantity to which the operator β belongs, will with certainty give the value b_i, the eigenvalue, if the system is described by the eigenfunction ψ_i (this is called a pure state). If the operators α and β commute, the simultaneous determination of the quantities corresponding to α and β will lead to exactly predictable values for both, namely the eigenvalues a_i and b_i.

Assume now that the wave function is not an eigenfunction of α; one can repeat the macroscopic arrangement of a given system and make repeatedly a certain measurement of an observable quantity represented by the operator α. Then the average value of the quantity should be calculated according to the following rule:

$$\bar{\alpha} = \int_\tau U^*\alpha U d\tau \tag{4-2}$$

where U is the wave function that describes the macroscopic state of the system at any given time. Note that Eq. 4-2 tells nothing about the result of a single measurement but represents an expected mean in the usual sense, i.e. if $a_1, a_2, a_3, \ldots, a_n$ are, respectively, the results of n measurements of the quantity α then

$$\bar{\alpha} = \frac{1}{n}\sum_{i=1}^{n} a_i$$

The order of writing the operator between U^* and U must be observed as the position matters for differential operators. If $U(q)$ is expanded into a complete set of orthogonal functions

$$U = \sum b_i\psi_i$$

Eq. 4-2 becomes

$$\bar{\alpha} = \int_\tau \sum b_i^*\psi_i^*\alpha\sum b_i\psi_i d\tau = \sum\sum b_i^*b_j \int \psi_i^*\alpha\psi_j d\tau \tag{4-3}$$

Now if ψ_i is an eigenfunction of the operator α, then

$$\alpha\psi_i = a_i\psi_i$$

and Eq. 4-3 becomes

$$\bar{\alpha} = \sum b_i^* b_i a_i$$

where one can interpret $b_i^*b_i$ in the following way: If a system is in a state described by the function U, which can be expanded as above into a series of eigenfunctions of α, and a single measurement of the quantity α is made, then $b_i^* b_i$ is the probability that the measurement will yield the value a_i. Thus if the function U is one of the eigenfunctions ψ_i then all the coefficients are identically zero except b_i and $b_i^* b_i = 1$. Then every measurement will give precisely the value a_i just as in Eq. 4-1. Thus the formulation of quantum mechanics is based on three fundamental postulates:

1. To every observable quantity there is assigned an operator.
2. If ψ_i is an eigenfunction of the operator α, i.e.

$$\alpha\psi_i = a_i\psi_i$$

   a single measurement will yield an exactly predictable value of $\alpha = a_i$.
3. If a system is described by a well-behaved function U, then the expected mean of a large number of measurements is given by $\bar{\alpha} = \int U^* \alpha U d\tau$.

These postulates will now be further elucidated by application to some simple systems.

B,5. The Schrödinger Equation. Assume a system with n degrees of freedom, then $U(q_1, q_2, \ldots, q_n, t)$ describes the system at any time t, and $H(p, q, t)$[3] is the Hamiltonian operator associated with the system according to the above rules. Then $U(q, t)$ is a solution of the following differential equation

$$HU(q, t) = -\frac{h}{2\pi i}\frac{\partial}{\partial t} U(q, t) \tag{5-1}$$

where

$$\int_\tau U^* U d\tau = 1$$

This is the time dependent Schrödinger equation and is exact[4] as long as one excludes velocities which approach the velocity of light. This equation can be written in general[5] as

$$\sum_{j=1}^{n} \nabla_j^2 U - \frac{8\pi^2 M}{h^2} V(q, t) U = \frac{4\pi M}{ih}\frac{\partial}{\partial t} U(q, t) \tag{5-2}$$

where all masses M are assumed equal.

[3] $H(p, q, t)$ is a short way of writing $H(p_1, p_2, \ldots, p_n; q_1, q_2, \ldots, q_n, t)$.

[4] One must also exclude certain problems in electrodynamics.

[5] There exists a second equation for the conjugate complex function U^* with $-i$ instead of $+i$, i.e.

$$\sum \nabla^2 U^* - \frac{8\pi^2 M}{h^2} V(q, t) U^* = -\frac{4\pi M}{ih}\frac{\partial}{\partial t} U^*$$

This is a second order linear partial differential equation and can be treated by the ordinary methods of solving boundary-value problems. This is sometimes called the wave equation, because it has the same form[6] as a wave equation in a medium in which the refractive index varies from place to place.

If one assumes that V, the potential energy, is a function of the coordinates only, then an elementary solution of Eq. 5-2 can be written:

$$U(q, t) = \psi(q)f(t) \tag{5-3}$$

Substituting Eq. 5-3 in Eq. 5-2 and separating variables in the usual way one obtains two total differential equations

$$\frac{1}{f}\frac{df}{dt} = -\frac{Lih}{4\pi M} \tag{5-4}$$

$$\sum \nabla_j^2\psi - \frac{8\pi^2 M}{h^2} V\psi = -L\psi \tag{5-5}$$

$$f(t) = e^{-\frac{ihL}{4\pi M}t}$$

Now calculating the average value of the energy, $\bar{E}$, according to the rule of Eq. 4-2

$$\bar{E} = \int_\tau U^* H U d\tau$$

$$\bar{E} = \int_\tau U^*\left(-\frac{h}{2\pi i}\frac{\partial}{\partial t}U\right)d\tau = -\frac{h}{2\pi i}\int f^*\frac{df}{dt}\psi^*\psi d\tau$$

But

$$f = e^{-\frac{ihL}{4\pi M}t}$$

$$\bar{E} = -\frac{h}{2\pi i}\left(-\frac{ihL}{4\pi M}\right) = \frac{h^2}{8\pi^2 M}L \tag{5-6}$$

Therefore the total energy is a constant, the time dependent Schrödinger equation, Eq. 5-2, has an oscillatory solution, and the frequency is E/h. The equation resulting from Eq. 5-5 by the elimination of L through Eq. 5-6 is called the time independent Schrödinger equation. If this is applied to the special case of a particle of mass m, then Eq. 5-5 becomes

$$\nabla^2\psi + \frac{8\pi^2 m}{h^2}(E - V)\psi = 0 \tag{5-7}$$

[6] Those familiar with the wave equations of classical physics will notice that this has the parabolic form in the place of the usual hyperbolic form, i.e. $\partial u/\partial t$ instead of $\partial^2 u/\partial t^2$. However, i is already included as an explicit factor in front of the time derivative so this is really a wave equation and not a diffusion equation.

The form of ψ is determined by the boundary conditions of the problem and it must be well behaved throughout configuration space. This will be true for only certain values of the energy E; these values may form a discrete set, a continuous set, or both. If the values of E form a discrete set of n values, one says there are n stationary states of the particle and the general wave function ψ can be written as a linear combination of elementary solutions

$$U = \sum_{i=1}^{n} a_i \phi_i e^{-\frac{2\pi i E_i}{h}t} \tag{5-8}$$

where ϕ_i is the eigenfunction associated with the eigenvalue E_i. This means that the system consists of a mixture of states; $a_i^* a_i$ tells how often the particle is in the state i with energy E_i. Usually E_i can be conveniently associated with a series of numbers n called total quantum numbers. Thus the quantum conditions that were arbitrarily imposed on classical mechanics in the old quantum theory arise logically as restrictions that are necessary to obtain well-behaved solutions of the Schrödinger equation in quantum mechanics.

B,6. Perturbation Theory. Assume a system in a stationary state, described by a time independent Schrödinger equation

$$(H - E)\Psi = 0 \tag{6-1}$$

The wave function ψ is determined if, besides the differential equation, the boundary conditions are given, e.g. for an atom, $\psi = 0$ at infinity; or for a stream of electrons, a plane wave at infinity.

Assume then that $\psi_1, \psi_2, \ldots, \psi_n$ form a complete (possibly continuous) set of orthogonal functions, determined by an arbitrary differential or linear operator equation, but restricted by the fact that each ψ satisfies the same boundary conditions as Ψ. One can then always write, as a generalization of a Fourier development, the jth solution of Eq. 6-1 (see Eq. 3-9)

$$\Psi_j = \sum_n b_{jn}\psi_n \tag{6-2}$$

To find the coefficients, one introduces Eq. 6-2 in Eq. 6-1, multiplies the result with a ψ_s^* (s arbitrary), and integrates. One has then, as a result of the orthogonality of the ψ,

$$\sum_n b_n(H_{sn} - \delta_{sn}E) = 0 \tag{6-3}$$

with the definition

$$H_{sn} = \int \psi_s^* H \psi_n d\tau \tag{6-4}$$

In Eq. 6-3, the index j, the first index of the b's, has been left out, since the set of Eq. 6-3 (one for every s) has not one, but an infinite set of solutions, and each set corresponds to a different j (i.e. Ψ_j, E_j).

The set of infinite linear equations for the unknowns b_n is solvable only if the following (infinite) determinant disappears:

$$|H_{sn} - \delta_{sn}E| = 0 \tag{6-5}$$

(secular equation). Each of the roots E of this equation is a possible eigenvalue E_j of Eq. 6-1, and the corresponding b_n's are the coefficients b_{jn} for Eq. 6-2.

To make the series (6-2) well convergent, and the approximate solution of Eq. 6-5 possible, one chooses as development functions ψ solutions of a "neighboring" problem

$$(H_0 - E_n^0)\psi_n = 0 \tag{6-6}$$

The difference,

$$H' = H - H_0 \tag{6-7}$$

is called the perturbation function; it is assumed that all the matrix elements H'_{sn} are small compared with the corresponding H_{sn}. One finds under these conditions, if one develops according to powers of H'_{sn}

$$b_{jn} = \frac{H'_{nj}}{E_j^0 - E_n^0} + \frac{1}{E_j^0 - E_n^0}\left[\sum_s \frac{H'_{ns}H'_{sj}}{E_j^0 - E_s^0} - \frac{(H'_{jj} - H'_{nn})H'_{nj}}{E_j^0 - E_n^0}\right], \quad j \neq n \neq s \tag{6-8}$$

$$b_{jj} = 1 - \frac{1}{2}\sum \frac{H'_{js}H'_{sj}}{(E_j^0 - E_s^0)^2} \tag{6-9}$$

$$E_j = E_j^0 + H_{jj} + \sum \frac{H'_{js}H'_{sj}}{E_j^0 - E_s^0} \tag{6-10}$$

Formulas (6-8) to (6-10) may also be found by writing

$$E = \int \Psi^* H \Psi d\tau = \sum_s \sum_n b_s^* H_{sn} b_s \tag{6-11}$$

and minimizing this expression under the condition that Ψ be normalized, i.e.

$$\sum b_n^* b_n = 1$$

These procedures are valid only in nondegenerate cases (the details for degenerate cases are not given here). The closer the ψ' are to the Ψ (i.e. the smaller H'_{sn}) the better the convergence.

B,7. Transitions. There are, in principle, two methods of treating transitions between quantum states; one considers a stationary system, the other considers a time-dependent system.

The former may be used in the investigation of transitions in a molecule induced by collisions with a stream of atoms. Since a transition in the molecule produces a change in the velocity of the atom, an investigation of how many scattered atoms have had their velocity decreased reveals how many collisions have resulted in an internal transition of the molecule to a higher quantum state, while the number of scattered atoms with higher velocity reveals the number of transitions to lower states. The details of such a calculation are given in H,28 and H,29. For transitions induced by light this method would not work, since in the stationary state there are as many excitations as deexcitations and therefore the number of scattered light quanta—the only quantity to be measured, since their frequency is not changed—is not directly affected by these processes. One has therefore to use the method of time dependent perturbation, which is more complicated than that of stationary perturbation. (It could also be used in atomic collisions.)

In the method now considered, one uses the time dependent Schrödinger equation and permits H' (but not H_0) to depend explicitly on time.

One now makes the development not with the time independent but with the time dependent solutions U of Eq. 5-1 and assumes that the coefficients a vary slowly with time (i.e. slow compared to $e^{(2\pi i/h)Et}$).

One therefore writes

$$U_j = e^{-\frac{2\pi i}{h}E_jt}\Psi_j$$

$$= a_{jj}(t)e^{-\frac{2\pi i}{h}E_j{}^0t}\psi_j + \sum_{n\neq j} a_{jn}(t)e^{-\frac{2\pi i}{h}E_n{}^0t}\psi_n \tag{7-1}$$

as generalization of Eq. 5-8. This is inserted into Eq. 5-1

$$(H_0 + H')U = -\frac{h}{2\pi i}\frac{\partial U}{\partial t} \tag{7-2}$$

and leads to

$$\sum_n a_{jn}(t)e^{-\frac{2\pi i}{h}E_n{}^0t}H'\psi_n = -\frac{h}{2\pi i}\sum e^{-\frac{2\pi i}{h}E_n{}^0t}\psi_n\frac{\partial}{\partial t}a_{jn}(t) \tag{7-3}$$

Again multiplying with ψ_s^*, integrating, and omitting the first index j as before, one has,

$$\sum e^{-\frac{2\pi i}{h}(E_n{}^0 - E_s{}^0)t}H'_{sn}a_n(t) = -\frac{h}{2\pi i}\frac{\partial}{\partial t}a_s(t) \tag{7-4}$$

This is a set of linear differential equations for a_s. One can integrate them approximately by assuming that at $t = 0$ all the systems were in the state

$$\Psi_j = \psi_j$$

i.e.

$$a_j(0) = 1 \qquad a_n(0) = 0$$

and by using, for a short time interval, these initial values of the a's in the left side of Eq. 7-4. Then

$$-\frac{h}{2\pi i}\frac{\partial a_j}{\partial t} = H'_{jj}$$

This gives only a change of the exponent from

$$-\frac{2\pi i}{h}E_j^0 t \quad \text{to} \quad -\frac{2\pi i}{h}(E_j^0 + H'_{jj})t$$

in accordance with Eq. 6-10.

For $s \neq j$

$$-\frac{h}{2\pi i}\frac{\partial a_s}{\partial t} = e^{-\frac{2\pi i}{h}(E_j^0 - E_s^0)t}H'_{sn} \tag{7-5}$$

The next step may be made on different levels of rigor. In the more rigorous case, one considers the system as made up of molecules and photons, and the transition as consisting in the excitation of the molecule and the simultaneous disappearance of a photon. On a less rigorous level, one introduces a potential energy of interaction between the electrical field, $\Xi_0 e^{2\pi i\nu t}$, and the charge displacement er (this would not be legitimate for hard X rays), and writes

$$H' = e(r\Xi_0)e^{-2\pi i\nu t} \tag{7-6}$$

It turns out that, to get reasonable results, one has to assume that the incoming radiation has $\nu \sim (W_j^0 - W_s^0)/h$, but is not quite monochromatic. (If it were monochromatic, one would have to wait, according to Heisenberg's uncertainty principle, an infinitely long time, and the question as to what is the transition probability would lose any meaning.)

If one then performs the calculation, one finds

$$|a_s^2| = \frac{8\pi^3}{3hc}|M_\Xi^2|I_\nu t \tag{7-7}$$

where I_ν is the amount of energy flowing per second through unit area within unit frequency "band width" and M is the component of the electrical moment in the direction of the field

$$M_{js} = e\int \psi_j^* r_\Xi \psi_s d\tau \tag{7-8}$$

$|a|_s^2$ is the number of particles found in states s at time t, when at zero time all were in state j. This quantity is seen to be proportional to t (this is only valid for times so short that only a small fraction has made the transition; otherwise the approximation made following Eq. 7-4 is not permissible). Therefore

$$\frac{|a_s|^2}{t} = \frac{8\pi^3}{3hc}|M_x^2|I_\nu \tag{7-9}$$

is the transition probability, i.e. the fraction of particles making the transition in unit time.

A consideration of temperature equilibrium in the field of black body radiation (there must be an equal number of particles per second excited by light absorption as are returning to the lower state through spontaneous emission of a quantum) shows that the probability of spontaneous emission is given by

$$\frac{64\pi^4\nu^3}{3hc^3}|M^2| \tag{7-10}$$

The value of M is determined by the initial wave function ψ_j and the final wave function ψ_s. In order that M be different from zero (the transition be allowed or be a dipole transition), the angular momenta of states j and s may differ by only 1 or 0 (this is a necessary, not a sufficient, condition). But even if M is zero, other transitions may be possible, although with much lower probability. In Eq. 7-6, the system (atom or molecule) was considered negligibly small compared to the wavelength of the light λ, i.e. phase differences between different parts of the atom or molecule were neglected. If the light is propagated in the z direction, Eq. 7-6 must be multiplied by a factor

$$e^{2\pi i\frac{z}{\lambda}} = 1 + 2\pi i\frac{z}{\lambda} - \frac{1}{2}\left(2\pi\frac{z}{\lambda}\right)^2 \cdots$$

In the previous calculation, only the first term was considered. The second gives a contribution

$$\frac{2\pi i}{\lambda}M'$$

where M' is called a quadripole moment and is given by

$$M' = \int \psi_s^* z r_x \psi_j d\tau$$

The resultant quadripole transition probability is found from Eq. 7-10 by replacing $|M^2|$ by $(4\pi^2/\lambda^2)\ |(M')^2|$ and stands to that for a dipole transition roughly in the ratio

$$\left(2\pi\frac{\text{dimension of atom}}{\lambda}\right)^2 : 1$$

B,8. Application of Quantum Mechanics to Simple Systems. The time independent Schrödinger equation will now be solved for a few simple systems. Since the Hamiltonian operator is invariant under transformation of coordinates, the coordinate system that is most convenient for the problem will be chosen. Consider a particle of mass m, moving along the x axis under the influence of no forces, that is, $V = 0$, then the wave equation $H\psi = E\psi$ becomes

$$\frac{-h^2}{8\pi^2 m}\frac{d^2\psi}{dx^2} = E_x\psi \tag{8-1}$$

$$\psi = Ae^{\frac{2\pi i}{h}\sqrt{2mE_x}\,x} + Be^{-\frac{2\pi i}{h}\sqrt{2mE_x}\,x} \tag{8-2}$$

In order for ψ to be finite for all values of x, $\sqrt{2mE_x}$ must be real, therefore the only restriction is that E_x must be positive. Thus translational motion is not quantized and E_x may vary continuously from 0 to $+\infty$. A and B are arbitrary constants. Since $Ae^{(2\pi i/h)\sqrt{2mE_x}\,x}$ and $Be^{-(2\pi i/h)\sqrt{2mE_x}\,x}$ are linearly independent when $E \neq 0$, the energy spectrum for a free particle is degenerate. These eigenfunctions are not quadratically integrable and thus the problem of normalization presents greater difficulties than in other problems and is not discussed here, but can be found in any one of the texts on quantum mechanics listed in the bibliography. If one uses the wave functions from Eq. 8-2 to calculate the linear momentum according to Eq. 4-2

$$p = \int U^* \frac{h}{2\pi i}\frac{\partial}{\partial x} U d\tau$$

and uses the expression with positive component only or that with negative component only, one finds, in agreement with classical theory,

$$p = \pm\sqrt{2mE_x} \tag{8-3}$$

The wave functions represent plane waves, going to the left or right, with wavelength

$$\frac{h}{\sqrt{2mE}} = \frac{h}{p} \tag{8-4}$$

in agreement with de Broglie's assumption.

The simple harmonic oscillator in one dimension. A simple harmonic oscillator consists of a particle of mass m held to an equilibrium position Q by a restoring force $F = -k^2x$ where x is the displacement of the particle from Q. Newtonian mechanics says this particle will oscillate about Q with an arbitrary amplitude and energy with a frequency of

$$\nu = \frac{1}{2\pi}\sqrt{\frac{k^2}{m}}$$

If

$$V = \tfrac{1}{2}k^2x^2$$

and the origin of the coordinate system is placed at the point Q then $H\psi = E\psi$ becomes

$$\frac{d^2\psi}{dx^2} + \frac{8\pi^2 m}{h^2}\left(E - \frac{1}{2}k^2x^2\right)\psi = 0 \qquad (8\text{-}5)$$

Changing variables, let

$$y = \left(\frac{4\pi mk^2}{h^2}\right)^{-\frac{1}{4}} x$$

then if $\lambda = \dfrac{2E}{h\nu}$

$$\frac{d^2\psi}{dy^2} + (\lambda - y^2)\psi = 0 \qquad (8\text{-}6)$$

The problem is now one of finding the values of λ for which Eq. 8-6 has well-behaved solutions, i.e. ψ must decay exponentially as $y \to \infty$. Try a solution of the form

$$\psi = e^{-\frac{1}{2}y^2} f_n(y) \qquad (8\text{-}7)$$

Differentiating and substituting in Eq. 8-6 one obtains

$$\frac{d^2 f}{dy^2} - 2y\frac{df}{dy} + (\lambda - 1)f = 0 \qquad (8\text{-}8)$$

This differential equation has a solution in terms of a power series

$$f_n(y) = \sum_n a_n y^n \qquad (8\text{-}9)$$

This solution will cause the wave function (Eq. 8-7) to become infinite as y grows very large, consequently Eq. 8-6 will have well-behaved solutions (Eq. 8-7) only if the series breaks off after a finite number of terms which, it can be shown, requires

$$\lambda = 2n + 1 \quad \text{where } n = 0, 1, 2, \cdots, n$$

Therefore

$$2n + 1 = \frac{2E_n}{h\nu}$$

or

$$E_n = (n + \tfrac{1}{2})h\nu \qquad (8\text{-}10)$$

The functions $f_n(y)$ are called Hermite polynomials defined by

$$f_n(y) = (-1)^n e^{y^2} \frac{d^n}{dy^n} e^{-y^2} \qquad (8\text{-}11)$$

Thus quantum mechanically the linear oscillator can no longer have an arbitrary energy but has instead a set of discrete values separated by

$h\nu$ where ν is the classical frequency and furthermore the zero-point energy is $\frac{1}{2}h\nu$. This is necessary to satisfy the minimum condition $\Delta p \Delta q = h$. Thus even in the lowest energy state a simple harmonic oscillator still has an energy $\frac{1}{2}h\nu$. If one extends this to the n-dimensional oscillator as must be done in the case of polyatomic molecules then

$$\frac{E}{h} = \nu_1\left(n_1 + \frac{1}{2}\right) + \nu_2\left(n_2 + \frac{1}{2}\right) + \cdots + \nu_n\left(n_N + \frac{1}{2}\right) \quad (8\text{-}12)$$

where ν_n are the frequencies of the characteristic modes of vibration of the molecule.

The rigid rotator in space. Assume a particle of mass m which rotates about a point O so that the particle is always a distance r from the point. Then choose a spherical polar coordinate system (r, θ, ω) with origin at O. The Laplace operator in spherical coordinates is

$$\nabla^2 = \frac{1}{r^2}\frac{\partial}{\partial r}\left(r^2 \frac{\partial}{\partial r}\right) + \frac{1}{r^2}\frac{\partial}{\partial \mu}\left[(1 - \mu^2)\frac{\partial}{\partial \mu}\right] + \frac{1}{r^2(1 - \mu^2)}\frac{\partial^2}{\partial \omega^2} \quad (8\text{-}13)$$

where $\mu = \cos\theta$. If

$$U = e^{-2\pi i \frac{E}{h} t}\psi$$

and only the stationary states are of interest at this time, the Schrödinger equation for the rigid rotator (r = const) can be written

$$\frac{1}{r^2}\frac{\partial}{\partial \mu}\left[(1 - \mu^2)\frac{\partial \psi}{\partial \mu}\right] + \frac{1}{r^2}\frac{1}{1 - \mu^2}\frac{\partial^2 \psi}{\partial \omega^2} + \frac{8\pi^2 m}{h^2}E\psi = 0 \quad (8\text{-}14)$$

Making the substitution

$$\sigma^2 = \frac{8\pi^2 m r^2}{h^2}E$$

the equation becomes

$$\frac{\partial}{\partial \mu}\left[(1 - \mu^2)\frac{\partial \psi}{\partial \mu}\right] + \frac{1}{1 - \mu^2}\frac{\partial^2 \psi}{\partial \omega^2} + \sigma^2\psi = 0 \quad (8\text{-}15)$$

Eq. 8-15 can be solved in the usual way by separating variables and finding total differential equations which have well-behaved solutions in analogy to the procedure used for the simple harmonic oscillator. Thus one obtains for the orthogonal normalized solutions of Eq. 8-15

$$\psi_{l,m}(\mu, \omega) = \frac{(-1)^l}{2^l l!}\sqrt{\frac{2l+1}{2}\frac{(l - |m|)!}{(l + |m|)!}}\sin^{|m|}\theta \frac{d^{l+|m|}\sin^{2l}\theta}{d\mu^{l+|m|}}\frac{1}{\sqrt{2m}}e^{im\omega} \quad (8\text{-}16)$$

where $l = 0, 1, 2, \cdots$ and $m = 0, \pm 1, \pm 2, \cdots, \pm l$.

The energy values of the rigid rotator are given by

$$\sigma^2 = \frac{8\pi^2 mr^2 E}{h^2} = l(l+1) \tag{8-17}$$

therefore

$$E_l = l(l+1)\frac{h^2}{8\pi^2 mr^2} \tag{8-18}$$

Similarly, it can be shown that the total angular momentum given by

$$M^2 = M_x^2 + M_y^2 + M_z^2$$

can be calculated using the wave functions from Eq. 8-16. One obtains

$$M^2 = \int \psi_{l,m}^* M^2 \psi_{l,m} d\tau = l(l+1)\left(\frac{h}{2\pi}\right)^2 \tag{8-19}$$

Thus the square of the total angular momentum is also quantized and has predictable values given in Eq. 8-19 and can be determined simultaneously with the total energy. The physical meanings of the two quantum numbers l and m then are: l is the total angular momentum quantum number and m is the quantum number associated with different permitted orientations in space, i.e. values of the angle θ. If one writes the operator for the z component of the angular momentum in spherical polar coordinates it becomes $(h/2\pi i)(\partial/\partial\omega)$ and the eigenvalues of this operator can be shown to be $mh/2\pi$; the eigenfunctions are the same as the eigenfunctions for M^2 or H. Therefore not only is the square of the total angular momentum and consequently of the angular velocity of the rigid rotator allowed certain discrete values but also only certain orientations in space are possible.

The next system to be considered is the hydrogen atom. Assume a nucleus of mass m_1, a positive Ze charge,[7] and an electron ($-e$, mass m_2) a distance r from the nucleus under the influence of a Coulomb attraction $V = -Ze^2/r$. The most convenient coordinate system for a discussion of this problem is chosen as follows: Let x, y, z be the coordinates of the center of mass of the system and place a spherical coordinate system (r, θ, ω) at the center of mass. Then the coordinates of the nucleus will be

$$\left(\frac{m_2}{m_1 + m_2} r, \theta, \omega\right)$$

and the coordinates of the electron will be

$$\left(\frac{m_1}{m_1 + m_2} r, \theta, \omega\right)$$

[7] $Z = 1$ for hydrogen but to simplify later discussions it will be kept in the equation as Z.

In this coordinate system the classical Hamiltonian operator H can be written

$$H = \frac{m_1 + m_2}{2}\left[\left(\frac{dx}{dt}\right)^2 + \left(\frac{dy}{dt}\right)^2 + \left(\frac{dz}{dt}\right)^2\right] + \frac{m_1 m_2}{m_1 + m_2}$$

$$\left[\left(\frac{dr}{dt}\right)^2 + r^2\left(\frac{d\theta}{dt}\right)^2 + r^2 \sin^2\theta\left(\frac{d\omega}{dt}\right)^2\right] - \frac{Ze^2}{r} \quad (8\text{-}20)$$

and the quantum mechanic operator becomes

$$H = \frac{-h^2}{8\pi^2(m_1 + m_2)}\nabla^2(x, y, z) - \frac{h^2}{8\pi^2}\frac{(m_1 + m_2)}{m_1 m_2}\nabla^2(r, \theta, \omega) - \frac{Ze^2}{r} \quad (8\text{-}21)$$

So that Schrödinger's equation $H\psi = E\psi$ becomes

$$-\frac{h^2}{8\pi^2(m_1 + m_2)}\nabla^2(x, y, z)\psi - \frac{h^2}{8\pi^2\mu}\nabla^2(r, \theta, \omega)\psi - \frac{Ze^2}{r}\psi = E\psi \quad (8\text{-}22)$$

Let[8] $\psi = F(x, y, z)\psi(r, \theta, \omega)$ then in the same way as before, one obtains two equations, one of which is

$$\frac{h^2}{8\pi^2(m_1 + m_2)}\nabla^2(x, y, z)F(x, y, z) + (E - E^0)F = 0 \quad (8\text{-}23)$$

where E = total energy, E^0 = internal energy, and $E - E^0$ = translational energy of the center of mass. The other equation from Eq. 8-20 becomes

$$\frac{h^2}{8\pi^2\mu}\nabla^2(r, \theta, \omega)\psi + \left(E^0 + \frac{Ze^2}{r}\right)\psi = 0 \quad (8\text{-}24)$$

Eq. 8-23 is the Schrödinger equation of the three-dimensional translational motion of a mass equal to the total mass of the system located at the center of mass of the system. This equation is like Eq. 8-5 but it is now in three dimensions. The variables can be separated in the usual way and the wave functions determined. The results would be analogous to Eq. 8-7 and one would find that the translational energy of the system calculated in this way is exactly equal to the classical kinetic energy of translation of such a system, and is not quantized but may have all positive energies from zero to high values, as long as one does not approach the velocity of light. It is usually possible to separate out the translational degrees of freedom of any system in this way regardless of the complexity of the system.

Eq. 8-24 is the Schrödinger equation for the internal energy of the hydrogen atom. The internal energy of the hydrogen atom, E^0, must be negative if the electron is bound. This energy is quantized as is shown below; however, if E^0 is positive the energy is no longer quantized but takes on continuous values. This is experimentally observed in the

[8] The time factor has already been eliminated by separation of variables.

hydrogen spectrum where the spectrum consists of discrete lines and a continuum.

Eq. 8-24 can be solved in the usual way by separating variables and one will obtain an r-dependent equation and an equation exactly like Eq. 8-14 for the rigid rotator; then the solutions of Eq. 8-24 can be shown to be Eq. 8-16 multiplied by an r-dependent solution. One defines the following quantities

$$\rho = \frac{2Z}{na_0} r \quad \text{and} \quad a_0 = \frac{h^2}{4\pi^2 \mu e^2}$$

a_0 is the radius of the first Bohr orbit as calculated by quantum theory using classical mechanics as a basis.

One obtains for the orthogonal normalized wave functions of the hydrogen atom

$$\psi_{n,l,m} = \frac{1}{\sqrt{2\pi}} e^{im\omega} \left[\frac{(2l+1)(l-|m|)!}{2(l+|m|)!}\right]^{\frac{1}{2}} P_l^{|m|}(\mu)$$
$$\left\{\left(\frac{2Z}{na_0}\right)^3 \frac{(n-l-1)!}{2n[(n+l)!]^3}\right\}^{\frac{1}{2}} e^{-\rho/2}\rho^l L_{n+l}^{2l+1}(\rho) \qquad (8\text{-}25)$$

where $P_l^{|m|}(\mu)$ are the associated Legendre functions and $L_{n+l}^{2l+1}(\rho)$ are the associated Laguerre polynomials.

Just as in the case of the rigid rotator, l is the total angular momentum quantum number usually called the orbital angular momentum as $l(l+1)$ gives the possible values of the square of the momentum of the electron moving about the nucleus in stable orbits measured in units $(h/2\pi)^2$; m gives the allowed orientations of these orbits.

In order to obtain well-behaved solutions for the r-dependent equations it is necessary to select another hitherto undetermined constant, as an integer. This defines another quantum number, n', called the radial quantum number which can have values 0, 1, 2, . . . ; or the total quantum number is defined as $n = n' + l + 1$ where n appears explicitly in the wave function given in Eq. 8-25. Its physical significance becomes apparent if one uses the eigenfunctions $\psi_{n,l,m}$ to calculate the allowed energy values E_n^0, in the usual manner.

Thus having obtained the eigenfunctions of the Schrödinger equation of the hydrogen atom, one can calculate the allowed energies (eigenvalues) according to the equation

$$H\psi_{n,l,m} = E_n\psi_{n,l,m} \qquad (8\text{-}26)$$

and one obtains

$$-E_n = \frac{4\pi^2 Z^2 e^4}{n^2 h^2}\left(\frac{m_1 m_2}{m_1 + m_2}\right) \qquad (8\text{-}27)$$

Therefore the energy values are dependent only on the total quantum number n. The electronic energies given in Eq. 8-27 are the same energies as those calculated by the old Bohr theory and the energy differences for different values of n correspond exactly to the observed spectra of the hydrogen atom ($Z = 1$), that is,

$$E_n - E_m = \frac{-2\pi^2 e^2 m_1 m_2}{h^2(m_1 + m_2)}\left(\frac{1}{n^2} - \frac{1}{m^2}\right)$$

$$= Rhe\left(\frac{1}{n^2} - \frac{1}{m^2}\right) \tag{8-28}$$

This is the Balmer formula for observed spectral series; since the energy depends only upon n, these energy values are degenerate because one can obtain the same values of n for different values of l and m. Since $n = n' + l + 1$ and $m = \pm l, \pm(l-1), \pm(l-2), \cdots, 0$ for every value of l there are $2l + 1$ values of m that give the same value of n and $l = n - 1$, $n - 2, \cdots, 0$ so the degeneracy of the eigenvalues given in Eq. 8-27 is

$$\sum_{l=0}^{n-1} 2l + 1 = 1 + 3 + \cdots 2n - 1 = n(n-1) + n = n^2$$

This says that for a given value of n there are n^2 different states in which the electron can move with the same energy E_n. These states are characterized by different angular momenta (l) and different orientations in space (m). Usually m is called the magnetic quantum number because, if the atom is placed in a magnetic field, the energies of the electron in the different states characterized by m are no longer equal but depend on the orientation of the state with reference to the applied magnetic field; one says the energy of the magnetic field has split the degeneracy. If the Schrödinger equation is solved, including the term due to the applied magnetic field, the solutions of Eq. 8-24 contain m explicitly for different energy eigenvalues. This effect is observed experimentally and is called the Zeeman effect.

Thus the stable states of the electron in the hydrogen atom are completely described by the three quantum numbers n, l, and m and the following notation taken from spectroscopy is used. Electronic states for which $l = 0, 1, 2, 3$ are described by the letters s, p, d, and f, respectively, and the value of n is written in front of the letter to designate the state. Thus the electronic structure of the ground state of the hydrogen atom is given by $1s^1$ which is a shorthand way of saying one electron (superscript) is in the $n = 1$, $l = 0$ orbit. The notation is further illustrated below.

$n =$	1	2	2	3	3	3	$\cdots$ 4
$l =$	0	0	1	0	1	2	$\cdots$ 3
state $=$	$1s$	$2s$	$2p$	$3s$	$3p$	$3d$	$\cdots$ $4f$

B,9. Qualitative Discussion of Extension of Quantum Mechanics to Atoms Containing More Than One Electron. Although the hydrogen atom represents the limit of the problems for which the Schrödinger equation can be solved exactly, considerable qualitative understanding of more complicated systems can be obtained by analogy with the hydrogen atom.

For example, assume that a more complex atom consists of a nucleus with charge $+Ze$ surrounded by Z electrons moving in the field of the nucleus and attracted to the nucleus by a Coulomb potential. Then one treats the problem as if each electron moved independently of the other electrons in the atom and were influenced only by the average field due to the other electrons and the field due to the nucleus. With these simplifying assumptions, it is now possible to consider the motion of each electron moving in an average central field that can be characterized by an effective nuclear charge Z' and each electron will be in its own state characterized by a hydrogenlike wave function. The electronic eigenfunctions in more complicated systems can still be given in terms of the three quantum numbers n, l, and m, but two electrons with the same n value but different l values will now have different energies. This is because the electron with smaller angular momentum goes closer to the nucleus and the attractive potential is stronger there because the nucleus is less completely screened at this point by the other electrons. Thus for a given n, the states of lowest l have the lowest energy. The m degeneracy still exists in the absence of an external magnetic field.

Many of the properties of atoms can be qualitatively understood from a consideration of the form of the one-electron wave functions given in Eq. 8-25. $\psi^*\psi d\tau$ is the probability of finding the electron in a volume element $d\tau$. This can be interpreted either as the fraction of a chosen time interval that a point electron spends in $d\tau$ or considered as if the electron were smeared out into a diffuse cloud so that the charge density at any point is given by $e\psi^*\psi$ where e is the charge on the electron.

The r-dependent factor of the square of the wave function multiplied by $4\pi r^2$ gives the relative probability of finding the electron at a distance r from the nucleus. The angular distribution of the electrons in a given orbit must also be considered. The angular distribution function varies with the quantum numbers l and m and is independent of the value of n. When $l = 0$, $m = 0$ the electron is said to be in an ns state and will always have spherical symmetry as can be seen from Eq. 8-25. (See Fig. B,9(a).) A $2p$ electron belongs to an orbit designated by the following quantum numbers, $n = 2$, $l = 1$, and $m = -1$, 0, or $+1$. The three values of m represent three independent wave functions. One can represent these three $2p$ wave functions in three mutually perpendicular planes of a Cartesian coordinate system with the nucleus at the origin. The charge distribution of an electron in the p orbitals is best represented by the drawings in Fig. B,9(b). When all three $2p$ orbits are occupied they give a combined electron cloud of spherical symmetry. The charge distributions for other combinations of quantum numbers are given in Fig. B,9(c, d).

Each curve is plotted symmetrically on each side of the vertical axis and the three-dimensional picture of the angular distribution of charge

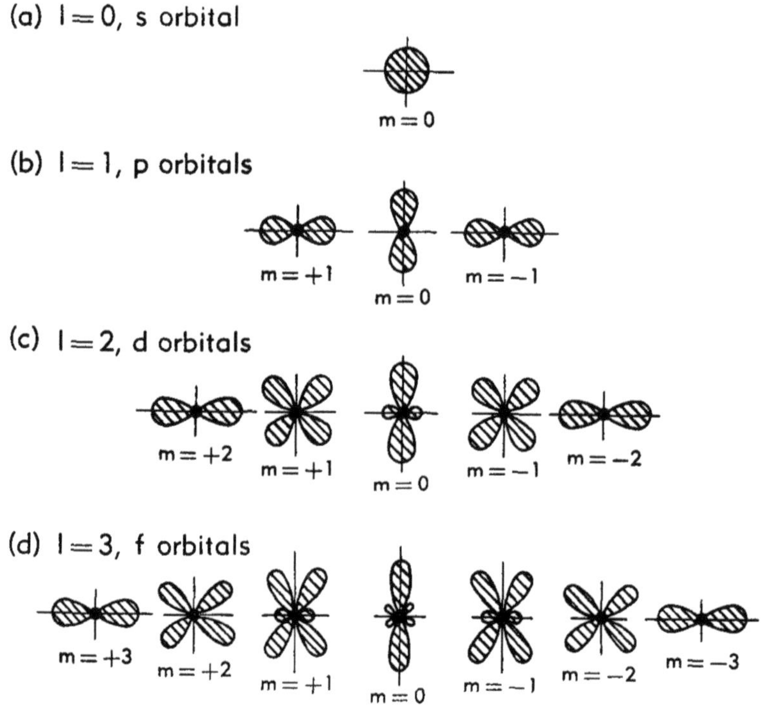

Fig. B,9. Angular distribution functions for different electronic states of hydrogenlike atoms. After [24] by permission.

can be represented by the solid of revolution formed by rotating each figure about its vertical axis. It should be emphasized that the electron is not confined to the shaded area in each figure but the length of the straight line joining the origin and any point on a given curve is a measure of the probability of finding the electron in the direction of that line. The importance of these wave functions becomes apparent in the discussion of molecules in Art. 12.

B,10. The Pauli Exclusion Principle and the Periodic Table. In the extension of quantum mechanics to describe atoms containing more than one electron, the mathematical problem becomes too complicated to obtain an exact solution of the Schrödinger equation. However, it is possible to obtain satisfactory quantitative laws for the motion of more than one electron in an atom by introducing an assumption about the distribution of electrons in the allowed orbits about the nucleus which will give agreement with the known experimental evidence. This principle is called the Pauli exclusion principle.

In its simplest form, this principle states that no two electrons can occupy orbits characterized by the same quantum numbers. This means that the necessary and sufficient difference between wave functions which must be fulfilled for the two wave functions to accommodate two electrons is that the two wave functions must be orthogonal. Consider an atom with atomic number Z. This atom consists of a nucleus with charge $+Z$ with the Z electrons moving in hydrogenlike orbits about the nucleus. The z electrons in the atom would thus be expected to fill successively the $1s$, $2s$, $2p_x$, $2p_y$, etc., states where the number of electrons that can simultaneously occupy a single state is restricted by the Pauli exclusion principle. Thus the state characterized by $n = 1$ would accommodate 1 electron and the state $n = 2$ would accommodate 4 electrons, i.e. $2s$, $2p(m = 1)$, $2p(m = -1)$, $2p(m = 0)$, etc. According to this scheme a periodic system of the elements could be built up in which hydrogen formed the first group, helium, lithium, beryllium, and boron would give a second group; but this differs from the well-known periodic classification of the elements developed by the chemists, if one expects that the chemical properties of atoms are due to the configuration of the electrons in the atoms. However, if one assumes that two electrons can occupy each hydrogenlike orbit, then one obtains the chemical periodic table. All electrons having the same value of the total quantum number n are said to belong to the same shell. Electrons with the same n and same l belong to the same subgroup. The periodicity of the chemical properties of elements is due to the fact that the chemical properties of an element are largely determined by the number of electrons in the outermost shell.

As the atoms get more complex, interactions between electrons become important, and some deviations in order from the simple picture occur; however, the correct periodic table can be constructed from spectroscopic data of neutral atoms and their ions (i.e. atoms from which electrons have been removed) along with the changes of electronic configuration in a strong magnetic field as determined by the Zeeman effect. Interpretation of spectroscopic data also requires that at most two electrons occupy the same orbit (and many orbits are actually occupied by two) if the orbits are defined by the three quantum numbers n, l, and m. However, if we define a fourth quantum number s, which can have only two values for electrons, then the Pauli principle is valid and states that no two electrons can have all four quantum numbers n, l, m, and s in common.

B,11. The Electron Spin. It remains to explain the reason for the additional quantum number s. The electron has an internal degree of freedom which must be quantized even after the position (n, l) and momentum (l, m) of the electron have been determined. This internal degree of freedom[9] is due to the angular momentum of the spin of the electron about its own axis. There are only two independent possibilities for s, namely, $\pm\frac{1}{2}$; thus the magnitude of the electronic spin angular momentum of an electron along an arbitrary axis is $\frac{1}{2}(h/2\pi)$. The spin does not greatly influence directly the energy of an electron moving in a given orbit. However, it is of great importance in determining the chemical properties of atoms and their combination with one another to form molecules.

B,12. Formation of Molecules. In principle, quantum mechanics enables one to write down the Schrödinger equation for any system of nuclei and electrons. The exact solution of this equation would then provide a complete description of the equilibrium configuration of the nuclei, the orbits and energies of the electrons, and in general the chemical and physical properties of the system. However, in practice, it is impossible to solve the Schrödinger equation except for a few simple idealized systems. Nevertheless a great deal of qualitative understanding of the combination of atoms to form stable molecules can be achieved by relatively simple considerations of the motion of electrons in the individual atoms as the atoms are moved close enough for the atomic wave functions to overlap. Since the empirical rules of valency, which had been introduced into chemistry to explain the combination of atoms to form molecules, had made possible the formulation of a periodic table almost exactly like the periodic table discussed in the last article on the basis of the electronic structure of the atoms, it was concluded that the electrons in the molecules can be divided into two separate sets, first those which continue to move in the field of the atomic nuclei as if each atom were isolated, and second, the outer shell (valence) electrons which are responsible for the formation of chemical bonds between two or more atoms. When two atoms A and B are brought sufficiently close to each other, the valence electrons from A and the valence electrons from B may interact in such a way as to counteract the Coulomb repulsive forces between the positively charged nuclei and form a stable molecule at some equilibrium position. The way in which electrons can interact to form stable molecules can be made plausible by a consideration of the form of the one-electron wave functions shown in Fig. B,9(a, b, c, and d).

[9] Uhlenbeck and Goudsmit [*4*] had introduced the idea of an intrinsic angular momentum of the electron into classical quantum theory and Stern and Gerlach [*2*, pp. 159–166] measured the magnitude of the magnetic moment due to spin.

A new assumption is introduced, known as the principle of maximum overlap[10], which can best be illustrated by considering its application to the water molecule [5, pp. 192–194]. Each hydrogen atom is in a $1s$ state; thus the distribution of the electron is spherically symmetric about each hydrogen nucleus. The oxygen atom has an electronic structure $1s^2$, $2s^2$, $2p^4$. The two $1s$ electrons form a closed shell and the two $2s$ electrons form a closed subgroup about the oxygen nucleus but the four $2p$ electrons are distributed as follows: two fill the $2p_z$ orbit, perpendicular to the plane of the paper and do not take part in the binding; one of the remaining electrons moves in the $2p_x$ and one in the $2p_y$ orbit shown in

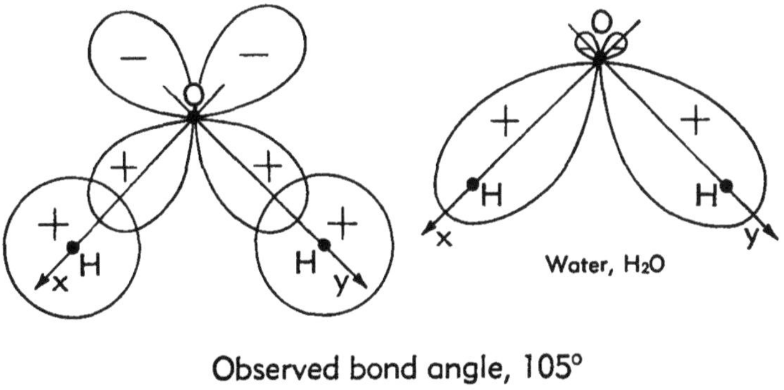

Fig. B,12. The water molecule. After [5] by permission.

Fig. B,12 and these alone take part in the binding of the water molecule. Now the principle of maximum overlap says the greatest binding energy and thus the most stable configuration is obtained by those electron arrangements which provide the largest negative charge between adjacent nuclei and thereby give the largest attraction. In the water molecule, this is obtained by placing one hydrogen atom along the x axis, and the other along the y axis so that there is maximum overlap of "electron clouds" as shown in Fig. B,12. Thus, the HOH angle in water should be about 90°. Experiment shows this angle to be 105° but here the mixing between the $2s$ and $2p$ orbits and the repulsion between hydrogen atoms has been neglected. Thus the directional properties of the wave functions shown in Fig. B,9(a, b, c, and d) are important and much qualitative information can be obtained about the size and shape of molecules by this type of consideration. Furthermore, the Pauli principle is still valid in molecular orbits formed by the interaction of two atomic orbits, so one can also predict in a qualitative way which combinations of atoms are allowed. For example, one can say two hydrogen atoms combine to form

[10] It is assumed that this is a qualitative description of the results of exact calculations which are too difficult to make.

a stable molecule because the two $1s$ wave functions in the two independent atoms combine to form a single wave function usually called a σ function which will accommodate two electrons, one with a spin $+\frac{1}{2}$ and one with a spin $-\frac{1}{2}$. On the other hand, two helium atoms do not combine because a single helium atom has two electrons in its $1s$ shell and so because of the Pauli principle, when two helium atoms are brought near to each other, neither can allow another electron in its orbit $n = 1$. Thus helium exists as a very stable monatomic substance.[11]

B,13. Structure of Molecules. Because of the complexity of the many-body problem, approximation methods had to be developed. To obtain any quantitative description of the energy states of molecules one important assumption is nearly always made which simplifies the problem considerably. Since the nuclei are relatively so much heavier than the electrons, the nuclear masses and the electron motions can be treated independently; so the motions of all the electrons are calculated assuming that the nuclei remain fixed in their equilibrium positions. The equilibrium positions and nuclear configuration can often be determined experimentally from spectra or by electron diffraction experiments. Then the motion of the nuclei is calculated assuming that they move in the field of an average steady state motion of the electrons. In this way one can write two independent Schrödinger equations, one for the electronic motions and one for the nuclear motions.

Two main approximation methods for the solution of the electronic motions of molecules have been developed. They are the Heitler-London-Slater-Pauling (HLSP)[12] method developed by the men for whom it is named and the molecular orbital (M.O.)[13] method of Lennard-Jones, Hund, Herzberg, and Mulliken.

It is beyond the scope of this book to give more than a qualitative idea of the quantum mechanics description of the structure of molecules.

The HLSP method starts with each electron moving in the spherically symmetric field of its own nucleus and then attempts to calculate the

[11] However, it is possible to form the He_2 molecule by exciting one electron into a $2s$ state; but this requires much energy.

[12] The HLSP approach has been widely used by chemists in formulating a theoretical basis for the chemical properties of compounds; [*6*] makes extensive use of this method. Craig [*7*] and Coulson [*8*] have used this method extensively in recent years but it is rapidly being replaced by the M.O. (molecular orbital) method.

[13] This method has been more successful in explaining the spectra of molecules but due to the greater mathematical complexity has been slow in developing. However, in the last five years due to the work of Mulliken and his coworkers [*9*] and independently of Lennard-Jones [*10*] and Coulson and his coworkers, extensive progress has been made in the development of the molecular orbital method. It appears that this method will be superior to the HLSP in calculating the properties of molecules making use of a minimum number of approximations and empirical facts. An excellent comparison of the M.O. and HLSP methods in their primitive form is given in [*11*].

interaction between the electrons as the nuclei are brought near to each other. Two definite electrons are assigned to each single bond, four to each double bond, etc. The M.O. method assumes that the nuclei are rigidly held in some equilibrium position and then feeds the electrons into the field of the entire molecule one by one. Every electron is considered as belonging, in principle, to the molecule as a whole. Up until the present time such drastic simplifications were necessary before the mathematical problem could be solved that only a few very simple molecules have been treated. However, present developments in the field give some promise that more theoretical calculations may be feasible in the near future (see [*7,8,9,10,12,13*]).

At present one must depend upon the spectra of polyatomic molecules and chemical properties to get even a roughly quantitative picture of their electronic structure. The electrons in molecules can be roughly ascribed to three classes. The most tightly bound electrons which form the inner shells in the free atoms are still essentially bound to a particular atomic nucleus and will take little or no part in binding the atoms in the molecule. If such an electron is excited, the resulting spectra will be like atomic spectra and will often be in the high energy (far ultraviolet) range of the spectrum. The next less energetic electrons are the so-called localized electrons which are shared between two nuclei but seldom move out of the range of these two nuclei. These electrons, if in pairs with opposite spin, are responsible for the single bonds between two atoms; but if these electron pairs have parallel spins, they may be antibonding in which case they will weaken bonds formed by other electrons between the same atoms. Finally, molecules contain weakly bound electrons which are sometimes called mobile electrons because they do not belong to any atom or isolated group of atoms but move through the entire molecule. These are analogous to the valence electrons in atoms and are responsible for some of the characteristic chemical properties of the molecule as well as the characteristic electronic spectra of many molecules.

The nuclear motions of molecules can be described in terms of the following Schrödinger equation

$$\sum_{i=1}^{N} \frac{1}{m_i} (\nabla_i^2 \psi) + \frac{8\pi^2}{h^2} [E - V]\psi = 0 \qquad (13\text{-}1)$$

where ψ is the wave function, E = the total energy, V = potential energy function averaged over the electron wave functions, m_i = mass of the nucleus, N = total number of atoms in the molecules, ∇_i^2 = Laplacian in Cartesian coordinates. The evaluation of $V(x_1, y_1, z_1, \ldots, x_N, y_N, z_N)$ by theoretical means requires a knowledge of the average field between adjacent nuclei due to the motions of the electrons. Thus far very little completely theoretical work has been done but the following treatment in combination with experiment has been one of the most reliable sources of information concerning the structure of polyatomic molecules. One assumes that the potential energy of the nuclei in the molecule is equal to zero for some equilibrium position for each atom. Then if one allows only small displacements of each nucleus from its equilibrium position one can treat the molecule as if it consisted of N oscillators tied to their equilibrium positions by springs with force constant k_{ii} and coupled to each other by springs with force constants k_{ij}. If one makes this assumption the mathematics is identical with the mathematics of coupled oscillators and it is usually possible to introduce $3N$ normal coordinates ξ_i in the place of the $3N$ Cartesian coordinates in Eq. 13-1. Furthermore, the wave function $\psi(x_1, y_1, z_1, \ldots)$ can be replaced by $\psi_1(\xi_1)\psi_2(\xi_2) \ldots \psi_i(\xi_i) \ldots \psi_{3N}(\xi_{3N})$. With these substitutions Eq. 13-1 can be resolved into a sum of $3N$ equations

$$\frac{1}{\psi_i}\frac{\partial^2\psi_i}{\partial\xi_i^2} + \frac{8\pi^2}{h^2}\left[E_i - \frac{1}{2}\lambda_i\xi_i^2\right] = 0 \qquad (13\text{-}2)$$

with $E = E_1 + E_2 \cdots E_i + \cdots E_{3N}$. But Eq. 13-2 is exactly like Eq. 8-5 of the simple harmonic oscillator. Thus the vibrational motion of the molecule may be considered to a good approximation as a superposition of simple harmonic motions in the $3N$ normal coordinates. The only difference between this and the classical oscillator problem is that the allowed energy values E_i are quantized and have a zero-point energy as was shown in the case of the oscillator. Thus

$$E = h\nu_1(n_1 + \tfrac{1}{2}) + h\nu_2(n_2 + \tfrac{1}{2}) \cdots h\nu_{3N}(n_{3N} + \tfrac{1}{2}) \qquad (13\text{-}3)$$

where ν_i's are the classical vibration frequencies and are the roots of the $3N$ equations of Eq. 13-2 solved simultaneously. For a nonlinear molecule six of these vibrations (for a linear molecule, five) are "nongenuine vibrations" (i.e. translations and rotations) and thus have a zero frequency and do not contribute to the vibrational energy. Thus there are $3N - 6$ genuine normal vibrations to be determined for each molecule ($3N - 5$ for linear molecules). These six (or five for linear molecules) coordinates can be separated out and one determines then another independent Schrödinger equation for the three translational degrees of freedom of the center of mass of the molecule; this will give an equation like Eq. 8-1 with continuous eigenvalues, i.e. continuous range of values of the translational energy of the molecules. These are of no interest in the structure of the molecule but are important in the treatment of the macroscopic properties of an aggregate of molecules such as discussed in statistical mechanics. Another equation can be written for the rotational motion of a polyatomic molecule. If the coordinates are properly chosen, the rotational motion can be represented by three equations (two, for linear molecules) like Eq. 8-14 where the three mutually perpendicular axes are chosen so that the moment of inertia (I_A) about one is a maximum, the moment of inertia about another (I_C) is a minimum and the moment of inertia about a third (I_B) is intermediate between I_A and I_C. (These are the principal axes of a rigid body.) For a linear molecule the rotational energy is given by Eq. 8-18

$$E_r = \frac{h^2}{8\pi^2 I}[j(j+1)] \tag{13-4}$$

where I is the moment of inertia about a point through the center of mass of the molecule and perpendicular to the molecule. The energy terms are a little more complicated for other more complicated molecules but have been solved and can be found in [*14*].

This is only a first approximation and in practice there is often interaction between rotation and vibration, but a large number of molecules have been studied and even these interactions and anharmonicities can be taken into account for a particular molecule. The rotational frequencies can be measured in the rotational infrared spectrum for non-symmetric molecules. These experimental values combined with the calculation of the principal axes usually enable one to get a pretty good idea of the interatomic distances. Recent measurements in microwave spectroscopy [*15*] have given a very accurate means of determining interatomic distances in many molecules.

The vibrational frequencies are more complicated. One is entirely dependent on experiment to determine the frequencies or the potential function. All of the fundamental frequencies do not always occur in the Raman and infrared spectra; there are more force constants in the potential energy equation than there are frequencies, so it is impossible to determine the potential energy function even when all the frequencies are known. However, there are often only a few possible structures consistent with the known chemical composition of a given molecule. Molecules with the same symmetry have similar Raman and infrared spectra. If the geometric configuration of a molecule is known it is possible to calculate the forms of the normal vibrations, and predict which one will occur in the Raman and which ones will occur in the infrared. For example, if the same frequency occurs in both the Raman and infrared spectra, the molecule can have no center of symmetry. Thus by a combination of calculations, experiments, and judicious guessing based on experience, one can usually determine the probable configuration of any given molecule that does not contain too many atoms.

It should be observed that most of the experimental methods, as well as the simplifying assumptions which make these methods useful in determining the structure of molecules, are restricted to the vapor state

Thus, the molecular energy levels can be divided into rotational, vibrational, and electronic levels whose relative order of magnitude can be roughly estimated in the following way. Suppose the molecule has a length of the order l. Therefore if an electron moves throughout an entire molecule with a characteristic length l then

$$\Delta p l \sim \frac{h}{2\pi}$$

$$\Delta p \sim \frac{h}{2\pi l}$$

$$E_e \sim \frac{\overline{\Delta p}^2}{2m} \sim \frac{h^2}{8\pi^2 l^2 m} \tag{13-5}$$

For $l =$ a few angstroms, $E_e \sim$ transitions in the visible and ultraviolet spectrum. The energy associated with a normal vibration of a molecule can be estimated using Eq. 8-10

$$E_v = h\left(n + \frac{1}{2}\right)\nu \sim h\nu$$

but $\nu = \sqrt{K/M}$ where K is the stiffness constant and M is of the order of the mass of a nucleus. If the displacement of the nuclei in a vibration were of the order of magnitude l, the electronic wave function would be seriously distorted and this would require

$$E_e \sim K_0 l^2 \quad \text{or} \quad K_0 \sim \frac{E_e}{l^2} \tag{13-6}$$

Thus

$$E_v \sim h\sqrt{\frac{E_e}{Ml^2}} \sim \left(\frac{m}{M}\right)^{\frac{1}{2}} E_e \tag{13-7}$$

The rotational energy can be estimated from Eq. 13-4 and 13-5:

$$E_r \sim \frac{h^2}{8\pi M a^2} \sim \frac{m}{M} E_r, \quad \left(\frac{m}{M}\right)^{\frac{1}{2}} \sim 10^{-2} \quad \text{and} \quad \frac{m}{M} \sim 10^{-4}$$

thus the vibrational energy is about $\frac{1}{100}$ of the electronic energy and the rotational energy is about $\frac{1}{100}$ of the vibrational energy. It is this different order of magnitude of the energies that makes it possible to consider the different internal degrees of freedom separately. However, the electronic spectra show not only electronic states but have superposed on the electronic states, the vibrational and rotational energies. This is the reason why the electronic spectra of molecules are wide bands while the electronic spectra of atoms which have no rotational or vibrational degrees of freedom are sharp line spectra.

CHAPTER 2. BOND ENERGIES

KARL F. HERZFELD
VIRGINIA GRIFFING

B,14. Introduction. Spectroscopic and Thermal Methods. With the knowledge that the atoms consist of positively charged nuclei surrounded by electrons it was first believed that molecules were held together by simple electrostatic forces between positively and negatively charged ions (polar bonds). It was assumed that certain elements had a tendency to gain extra electrons while others lost electrons easily and these combined to form stable molecules. Although this picture was adequate to explain the stability of many molecules, it gave no explanation for the existence of molecules consisting of two or more like atoms held together by strong chemical bonds (covalent bonds). It was in the development of quantum mechanics and the explanation of the periodic table of the elements that chemistry and physics merged their points of view in an attempt to understand more completely the structure of molecules. Quantum mechanics shows that chemical bond formation is the result of a profound rearrangement of electronic structure when two or more atoms are brought together, and gives a qualitative understanding of valency and bond formation; similarly, quantum mechanics has shown that molecules may exist only in certain permitted energy states and the transition from one to another is accompanied by the absorption or emission of energy. Ideally, a theory of molecular structure would enable one to determine the energy of formation of a molecule as a function of all variations in size and shape. Then, the coordinates for which this function has a minimum would give the stable configuration of the molecule. The difference between the value of this function at the minimum and the value for the atoms infinitely separated from one another would give the binding energy. Quantum mechanics, in principle, enables one to calculate this, but in practice, as has already been shown, the mathematical difficulties are too great and one must turn to experimental methods for the quantitative determination of bond energies and binding energies in molecules. The two primary sources of experimental data used in the determination of bond energies are spectroscopic and thermodynamic data. The sources of these data and their use in estimating bond energies are discussed in this chapter.

Thermal methods differ from spectroscopic methods in that thermochemical measurements are made on substances in bulk, at room temperature, or higher, and are then reduced to a standard temperature, usually 25°C. Since spectroscopy deals almost entirely with atoms and molecules in the free gaseous state, the energies determined in this way are at 0°K and zero pressure. Thermochemical measurements ordinarily involve matter in the liquid and solid state as well as the gaseous state. It is customary to reduce all thermochemical data to give energies for reactions between chemical substances in the "standard state." The "standard state" is defined as the state in which a substance normally exists at 25°C and one atmosphere pressure; e.g. nitrogen (N_2) and oxygen (O_2) are gases and water (H_2O) is a liquid. In some cases it is necessary to specify the allotropic form used as standard state if the substance exists in more than one form at the reference temperature. Carbon is one of the important substances for which a choice must be made as it exists as both diamond and graphite at 1 atmosphere and 25°C.[14] Thus in order to compare bond energies determined by thermochemical means with those determined from spectroscopic data, it is necessary to determine the energy needed to change a substance from a free gas at 0°K to its standard state at 25°C. Usually, spectroscopic values refer to the energy per molecule and must be multiplied by the Avogadro number to determine the energy per mole. Thermal measurements are usually made on arbitrarily chosen samples and must also be calculated for a mole. One can change the thermochemical value of the dissociation energy of a substance in its solid state at temperature T_1, to the value one would obtain for one mole of gaseous molecules at zero pressure and 0°K by the following thermodynamic relationship (see Sec. A)

$$H^{g}\,(0^\circ\mathrm{K}) = H^{s}_{T_1}\,(\text{experimental value}) + \int_{T_1}^{T_f} C^{s}_{p}dT + \int_{T_f}^{T_v} C^{l}_{p}dT + \int_{T_v}^{0^\circ\mathrm{K}} C^{g}_{p}dT + L_f + L_v \qquad (14\text{-}1)$$

where C^{s}_{p}, C^{l}_{p}, C^{g}_{p} are the molar specific heats at constant pressure for the solid, liquid, and gas, respectively; L_f and L_v are the latent heats of fusion and vaporization respectively and T_f and T_v are the freezing points and boiling points; $H^{s}_{T_1}$ is the experimentally determined heat of dissociation at constant pressure and temperature T_1.

Dissociation energies are large compared to specific heat terms; one need not have very accurate values of C_p but latent heats are of the same order of magnitude as the quantities being measured and thus accurate values are necessary when the reaction involves a change of phase. If the reaction does not involve a change of phase, one can use, for a change in H from 0° to 298°K, 5 kcal/mole for a monatomic gas and up to about 8 kcal/mole for a polyatomic gas. Since bond energies are of the order of 100 kcal/mole, the errors introduced by using these values are relatively insignificant. For this discussion, it is convenient to consider diatomic molecules and polyatomic molecules separately.

[14] References *[16]* are very useful compendiums of numerical results of thermochemical measurements and are invaluable in making thermochemical calculations.

B,15. Diatomic Molecules. In diatomic molecules, it is possible to give a rigorous definition of bond energy in terms of quantities that can often be determined experimentally. The bond energy of a diatomic molecule A—B is the energy that must be supplied to the molecule A—B in order to dissociate it completely into two neutral atoms A and B in their ground electronic states; both the molecule and the atoms should be at zero pressure and 0°K.

The most direct way of determining the bond energy is from the dissociation limits of the characteristic spectra of the molecule. The absorption spectrum is obtained when light possessing a continuous spectrum is passed through the substance, usually in the gaseous state. The molecule absorbs the characteristic frequencies of the molecule. The molecules may be excited by some other means, such as an electric discharge, and the excitation energy is emitted in the form of an emission spectrum. As has already been shown, the energy associated with the absorbed or emitted radiations can be obtained from the measured frequencies by the following relation

$$E_2 - E_1 = h\nu \tag{15-1}$$

The characteristic spectra of a diatomic molecule are simple band series which can be completely described in terms of three quantum numbers, namely, the rotational quantum number j, the vibrational quantum number v, and the electronic quantum number n. The bonding energy of a molecule is different for different electronic states of the molecule but at ordinary temperatures all the molecules are in the ground electronic state. Hence this is the state for which the bond energy is most important. The equilibrium internuclear distance can be determined from the rotational spectrum as shown in Chap. 1. The vibrational energy of a diatomic molecule can be represented in a potential energy curve such as that plotted in Fig. B,15 for H_2. If one assumes simple harmonic forces, the vibrational energy levels are given by

$$E = (n + \tfrac{1}{2})h\nu \tag{15-2}$$

However, as the molecule is stretched far from its equilibrium position, the frequency ν changes, the levels are no longer spaced equally, and Eq. 15-2 must be replaced by

$$E_n = (n + \tfrac{1}{2})h\nu - (n + \tfrac{1}{2})^2\alpha h\nu \tag{15-3}$$

where α is called the anharmonicity constant and must be experimentally determined from the vibration-rotation spectrum of the molecule. There have been many attempts to represent this type of potential energy-internuclear distance relationships by analytic expressions. The most satisfactory, because of its relative simplicity combined with a fairly good representation of the dependence of the potential energy of binding on the internuclear distance, is the following equation due to Morse:

$$V = D_e[1 - e^{-2\beta(r-r_e)}]^2 \tag{15-4}$$

where

$$\beta = \omega_e \sqrt{\frac{\pi^2 c\mu}{2hD_e}}$$

ω_e is the characteristic vibration frequency in cm^{-1}, μ is the reduced mass in atomic weight units, c and h are the velocity of light and Planck's constant. D_e is the dissociation energy measured from the minimum of

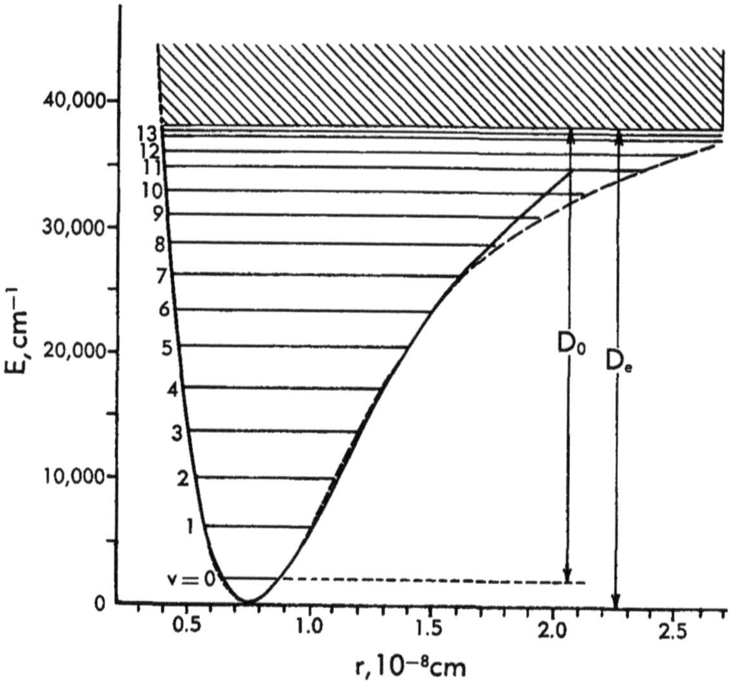

Fig. B,15. Potential curve of the H_2 ground state with vibrational levels and continuous term spectrum. The broken curve is the Morse curve. After [*17*] by permission.

the potential energy curve in cm^{-1}. In Fig. B,15 the full curve is drawn from spectroscopic data and the broken curve is the Morse curve. The cross-hatched space above the curve represents the continuous spectrum that results from the translational kinetic energy of the atoms which is no longer quantized once the bond is broken. Thus the dissociation energy is the energy necessary to just separate the molecule into free atoms and leave them with zero velocity at the asymptote of the converging vibrational levels. This is the same energy that was defined above as the bond energy. The spectra of over two hundred and fifty diatomic molecules have been studied and this material has been thoroughly reviewed and critically evaluated in two recently published books [*17,18*]. Although spectroscopic data are very important, it should be emphasized that they are often difficult to interpret and there is considerable disagreement among various observers for some important molecules. Thus according to recent evaluations [*17,18*] the dissociation energies of only about fifty diatomic molecules have been unambiguously determined.

The most direct thermochemical method of measuring the dissociation energy is to determine the equilibrium conditions for the dissociation

$$AB \rightleftarrows A + B$$

Then, the equilibrium constant K, which is a function of the temperature only, is given by

$$K = \frac{p_A p_B}{p_{AB}} \tag{15-5}$$

where p_A, p_B, and p_{AB} are the partial pressures of the three gases assumed ideal. If K is determined as a function of temperature, one can calculate the dissociation energy according to the following thermodynamic equation due to Van't Hoff (see Sec. A),

$$\frac{d \ln K}{dT} = -\frac{\Delta H}{R} \tag{15-6}$$

This value must then be converted to 0°K as discussed above.

The equilibrium constant, instead of its change with temperature, can be used directly for the determination of ΔH at 0°K if sufficient data are available. ΔH can be calculated according to the following equation from statistical mechanics

$$\ln K = \frac{\Delta H_0}{RT} + \ln \left[\mu_A^{3/2} \frac{\sigma}{10^{40} I} \left(\frac{T}{100} \right)^{5/2} \right] - \ln (1 - e^{-\theta/T}) + \ln \frac{\nu_A \nu_B}{\nu_{AB}} + 6.94 \tag{15-7}$$

where K is the equilibrium constant in atmospheres; μ_A is the reduced mass in atomic weight units; I is the moment of inertia of the molecule AB in g/cm²; $\theta = \omega hc/k$ is the characteristic temperature related to the fundamental vibration frequency; ν_A, ν_B, and ν_{AB} are the electronic statistical weights; σ is the symmetry number which is 2 for homonuclear diatomic molecules and 1 otherwise. I, θ, and the electronic statistical weights are determined from spectroscopic data. Eq. 15-6 and 15-7 both assume ideal gases; since these experiments are often conducted at high temperatures it is usually satisfactory to make this assumption. In addition, Eq. 15-7 assumes:

1. Temperature high enough to neglect effects of nuclear spin
2. Temperature high enough for rotational degrees of freedom to be treated classically
3. Neglect of interaction between vibrational and rotational degrees of freedom
4. The molecule to be a harmonic oscillator, i.e. anharmonic terms are neglected
5. The energy difference between the ground and first excited electronic state to be large compared to kT

The errors introduced by the first three assumptions are slight and Eq. 15-7 can be modified to take 4 and 5 into consideration if sufficient experimental data are available. The principles upon which Eq. 15-7 depends are discussed in the chapter on statistical mechanics (Chap. 5). In some cases, due to experimental difficulties, K may be the equilibrium constant for some reaction other than the direct dissociation of the molecule. In this case, Eq. 15-7 is replaced by a more complicated expression for the partition functions in order to determine the dissociation energy.

The dissociation energy of some diatomic molecule can be determined calorimetrically by combining thermochemical measurements of heats of formation with known values of other dissociation energies. For example, from the calorimetric determination of the heat of formation of NO from N_2 and O_2, combined with dissociation energy of N_2 and O_2 from spectroscopic measurements, one can calculate the dissociation energy of NO. The dissociation energy of NO is equal to the heat of formation of NO from N_2 and O_2, under standard conditions, minus the dissociation energies of N_2 and O_2. It is much simpler if all the substances are gases but this method can still be used for substances in the liquid or solid state if latent heats are known. For example, consider one of the alkali halides such as sodium chloride (NaCl). One would use the following Born-Fajans-Haber cycle:

$$\begin{array}{ccc} \mathrm{Na} + \mathrm{Cl} & \xleftarrow{\quad L_s(\mathrm{Na}) + \frac{1}{2}D(\mathrm{Cl_2}) \quad} & \mathrm{Na_{solid}} + \frac{1}{2}\mathrm{Cl_2} \\ D(\mathrm{NaCl}) \uparrow & & \downarrow Q(\text{formation NaCl}) \\ \mathrm{NaCl_{gas}} & \xleftarrow{\quad L_s(\mathrm{NaCl}) \quad} & \mathrm{NaCl_{solid}} \end{array}$$

where $L_s(X)$ means heat of sublimation of atoms or molecules X, and Q is a measured heat of formation; D(NaCl) is the dissociation energy of NaCl. Thus

$$D(\mathrm{NaCl})_{gas} = L_s(\mathrm{Na}) + \tfrac{1}{2}D(\mathrm{Cl_2}) + Q(\mathrm{NaCl}) - L_s(\mathrm{NaCl})$$

This value would, of course, have to be reduced to zero pressure and 0°K. This method is dependent on the principle (conservation of energy) that any arbitrarily chosen path of reaction which leads to dissociation must give the same value for the dissociation energy. In Table B,15 is given a list of dissociation energies of common diatomic molecules. In general, for diatomic molecules, when it is possible to determine the dissociation energy spectroscopically, this value is more accurate than dissociation energies of molecules determined by thermal methods. Of course, it is better if one can determine these dissociation energies by thermal and spectroscopic methods and use the thermal method simply as a check.

Table B,15

Molecule	*D, kcal*	*Molecule*	*D, kcal*
H—H	104.1	O—O	117.20
F—F	63.5	N—N	225.1*
Cl—Cl	58.1	C—O	256.1*
Br—Br	46.3	C—C	83*
I—I	36.4	C—N	175*
Li—Li	27.2	N—O	150
Na—Na	18.5	S—S	101
K—K	12.8		

* Doubtful.

B,16. Polyatomic Molecules. There are several different quantities which have in the past been called bond energies; some confusion has been caused by the failure to distinguish between them. Szwarc and Evans [*19*] have pointed out that two quantities should be considered, the bond dissociation energy and the average bond energy (the two are identical in diatomic molecules). In polyatomic molecules the bond dissociation energy is susceptible to direct experimental measurement, whereas the average bond energy is a calculated quantity obtained from the heat of atomization. This latter is the heat necessary to dissociate completely the molecule into the atoms, all in their electronic ground state, at 0°K.

This heat of atomization may be found in the following manner: Assume the tables give the heat of formation of a molecule A_2B from the elements as $-\Delta H^{(1)}$

$$(A_2B)_{gas} \rightarrow (A_2)_{gas} + (B)_{solid} - \Delta H^{(1)} \tag{16-1}$$

To atomize A_2B one has also to write

$$(A_2)_{gas} \rightarrow 2(A)_{gas} - \Delta H^{(2)} \tag{16-2}$$

$\Delta H^{(2)}$ is the heat of dissociation of gaseous A_2.

$$(B)_{solid} \rightarrow (B)_{gas} - \Delta H^{(3)} \tag{16-3}$$

$\Delta H^{(3)}$ is the heat of sublimation. Then

$$\text{heat of atomization} = \Delta H^{(1)} + \Delta H^{(2)} + \Delta H^{(3)} \tag{16-4}$$

While the heats of formation are often measured accurately, those of dissociation and particularly those of sublimation (particularly in the case of carbon) are often uncertain.

The bond dissociation energy may be defined for a bond between two fragments of a polyatomic molecule M—N in an analogous fashion to the dissociation energy of a diatomic molecule. Thus the energy in the bond between the fragment M and the fragment N is the energy necessary to separate the two fragments if one assumes that the original molecule MN and the fragments M and N are in the gaseous phase at zero pressure and 0°K. However, during this process, other bonds may be weakened or strengthened, new bonds may be formed, and electronic states of the atoms may be changed. All these effects are included in the bond dissociation energy.

The sum of the bond dissociation energies must be equal to the heat of atomization of the molecule. This relationship between these quantities can be best illustrated by an example. Consider the water molecule which has a heat of atomization per mole of 218 kcal. Since the reaction

$$H_2O \rightarrow O + 2H$$

requires the breaking of two O—H bonds, the "average bond energy" (see below) of the O—H bond in water is 109 kcal/mole. However, the bond dissociation energy of the first O—H bond in water is 118 kcal/mole, i.e.

$$H_2O \rightarrow H + OH - [D(H\text{—}OH) = 118 \text{ kcal/mole}]$$
$$OH \rightarrow O + H - [D(O\text{—}H) = 100 \text{ kcal/mole}]$$

This means that the removal of the first H atom has weakened the remaining OH bond. Thus it is obvious that like bonds do not have the same dissociation energy but are dependent on the fragments of the dissociation process. In general in a molecule ABC (e.g. HCN) the bond energy for the AB bond, i.e. the process ABC → A + BC, may be D_1 that for the bond BC in that fragment, i.e. for BC → B + C, is D_2. D_2 may be different from D_3, the energy of the process ABC → AB + C, AB → A + B needs D_4, different from D_1. But

$$D_1 + D_2 = D_3 + D_4 = \text{heat of atomization}$$

if the atoms end up in the same state in both procedures.

The bond dissociation energy may in principle be determined as follows:

1. Spectroscopically. This, however, is in general impossible at present, due to the complexity of the spectra of polyatomic molecules.

2. Thermodynamically. If the equilibrium concentration (and its temperature dependence) of the fragments can be measured, then Eq. 15-6 applies here also.

3. Rate measurements. As shown in Art. 18, the temperature coefficient of a rate constant is determined by the so-called heat of activation which can be found from the measurement of the rate.

Furthermore, from the first law of thermodynamics

heat of activation (decomposition) − heat of activation (recombination) = bond dissociation energy

If one assumes that the heat of activation for the recombination is zero, then the bond dissociation energy is equal to the heat of activation of the decomposition. An example is given in a paper by Szwarc [*20*]. Szwarc studies the rate of decomposition of a molecule into two radicals. The principal experimental problem is to eliminate the radicals formed in the decomposition before they have time to recombine or initiate other reactions making the kinetics complicated. He solves this problem by using a large excess of toluene as a carrier gas. Many radicals react instantly with the toluene according to the equation

$$C_6H_5CH_3 + R \rightarrow C_6H_5CH_2 + RH \tag{16-5}$$

and resulting benzyl radicals survive long enough to escape from the zone of reaction, eventually forming dibenzyl. Thus the rate of the unimolecular decomposition $R_1R_2 \rightarrow R_1 + R_2$ can be measured by the rate of formation of either R_1H, R_2H, or dibenzyl. This technique has been used for measuring the heats of the following bond dissociations:

(i)	$C_6H_5CH_2CH_3 \rightarrow$	$C_6H_5CH_2 + CH_3$	(63.0 kcal/mole)
(ii)	$C_6H_5CH_2Br \rightarrow$	$C_6H_5CH_2 + Br$	(50.0 " ")
(iii)	$C_6H_5CH_3 \rightarrow$	$C_6H_5CH_2 + H$	(77.5 " ")
(iv)	$N_2H_4 \rightarrow$	$2NH_2$	(60 " ")

In the first three of these reactions the heats of formation of the species concerned, except the benzyl radical, are all known from other sources. Therefore, the heat of formation of the benzyl radical can be calculated from any one of the three bond dissociation energies, and the concordance of the results provides a valuable cross check on the measured bond dissociation energies.

Similarly, if the heat of formation of the radicals is known from an independent experiment, one can calculate the bond dissociation energy, if the heat of formation of the compound is also known, e.g. the dissociation energy of the N—N bond in hydrazine (N_2H_4) is 60 kcal. The heat of formation of gaseous hydrazine from the elements in their standard state is 22 kcal/mole; thus one obtains the heat of formation of NH_2 radical as 41 kcal/mole, i.e.

$$\begin{aligned} \Delta H_f(NH_2) &= \tfrac{1}{2}\{[\Delta H_f(N_2H_4) + D(NH_2{-}NH_2)]\} \\ &= 41 \text{ kcal/mole} \end{aligned} \tag{16-6}$$

Szwarc [*20*] shows how, by using this value for the heat of formation of hydrazine, the dissociation energy of hydrogen (H_2), and the heat of formation of NH_3, one can calculate the bond dissociation energy of the first N—H bond in NH_3. This bond energy can be calculated according to the following equation which is simply an application of the first law of thermodynamics:

$$D(\mathrm{NH_2{-}H}) = \Delta H_f(\mathrm{NH_2}) + \tfrac{1}{2}D(\mathrm{H{-}H}) - \Delta H_f(\mathrm{NH_3})$$

$$D(\mathrm{NH_2{-}H}) = 41 + 52 + 11 = 104 \text{ kcal/mole} \tag{16-7}$$

One can calculate, according to this method, bond dissociation energies even if the rate of breaking the bond cannot be measured directly.[15]

The relationships between the bond dissociation energy and the heat of reaction have been used ingeniously by Kistiakowsky and Van Artsdalen [*21*] to determine the bond dissociation energy of the first C—H bond in methane (CH_4). The principle involved is again an application of the first law of thermodynamics. Given a reaction of the type,

$$\mathrm{RA + B \rightarrow R + AB}$$

then

$$\sum \text{dissociation energies of bonds formed} - \sum \text{dissociation energies of bonds broken is equal to heat of reaction}$$

Thus

$$D(\mathrm{R{-}A}) = D(\mathrm{A{-}B}) - \text{heat of the reaction}$$

Kistiakowsky determined the heat of reaction:

$$\mathrm{CH_4 + Br \rightarrow CH_3 + HBr} - 16 \text{ kcal/mole}$$

at 453°K. Recalculating the value of ΔH to 0°K, he obtained $\Delta H_0^0 = 15$ kcal/mole. Then applying the above principle and using a value for $D(\mathrm{H{-}Br})$ as 85.8 kcal/mole he concluded:

$$D(\mathrm{CH_3{-}H}) = D(\mathrm{H{-}Br}) + 15 \text{ kcal/mole} = 101 \text{ kcal/mole}$$

The average bond energy is most directly defined for molecules which have only one kind of bond, e.g. in a molecule AX_n where all the atoms X are in symmetric positions, the heat of atomization refers to the process:

$$\mathrm{AX}_n \rightarrow \mathrm{A} + n\mathrm{X}$$

and is set equal to n times the average energy of the A—X bond. Or the heat of atomization of $H_3C{-}(CH_2)_2{-}CH_3$ is set equal to 3 times the average bond energy of a C—C bond plus 10 times the average bond energy of a C—H bond.

The calculation of the average bond energy is therefore based entirely on the experimental determination of the heat of atomization, discussed previously.

[15] Other examples of the measurement of bond dissociation energies are discussed in [*20*].

The average bond energy is usually not equal to the bond dissociation energy of any particular bond in the molecule as exemplified in the above by H_2O. The average bond energy would be the energy required to rupture an A—X bond in an idealized experiment in which all the other bonds are extended to infinity while the molecule maintains its original geometric form. However, the sum of the average bond energies and the sum of the bond dissociation energies are equal to each other and are equal to the heats of atomization of the molecule.

The concept of average bond energy is particularly useful for groups of molecules for which one can assume (1) that the bond energy of a given bond is the same in all molecules in which the same combination of atoms appears and (2) that bond energies are additive.

The validity of these assumptions can be experimentally established for molecules which contain only normal single bonds, i.e. a bond localized between two atoms in which the wave functions of two-paired electrons[16] —one from each atom—thoroughly overlap. To find the average bond energies in molecules which contain two different kinds of bonds, one has to assume that these have the same values in at least two molecules of a homologous series, e.g.

heat of atomization of C_2H_6 = 1 average bond energy of a C—C bond plus 6 average bond energies of C—H bond

heat of atomization of C_3H_8 = 2 average bond energies of C—C bonds plus 8 average bond energies of C—H bond

The assumption may be tested by using the data for more than two molecules and has been confirmed by Prosen, Johnson, and Rossini [*22*]. The increment per CH_2 group becomes constant after the second or third member of a number of homologous series.

Even for single bonds there are significant deviations from the constant value for a given bond if that bond exists in a markedly different environment such that the angles are different from the normal configuration. In such cases, also, the bond lengths differ appreciably. This is particularly evident in the consideration of isomers, which are molecules containing the same number of like bonds but the chains are now branched rather than straight, e.g. the experimentally determined heat of formation of butane,

$$
\begin{array}{ccccc}
 & H & H & H & H \\
 & | & | & | & | \\
H-\!\!&C\!\!&-\!\!C\!\!&-\!\!C\!\!&-\!\!C-H \\
 & | & | & | & | \\
 & H & H & H & H
\end{array}
$$

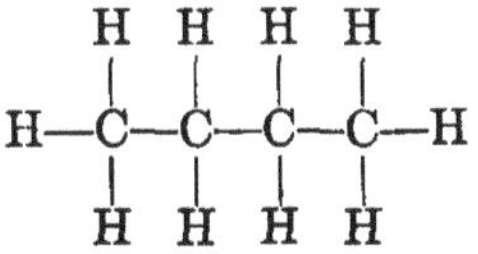

[16] If the resultant spin of two electrons is zero, they are called "paired electrons."

is 1044.8 kcal/mole while the heat of formation of isobutane

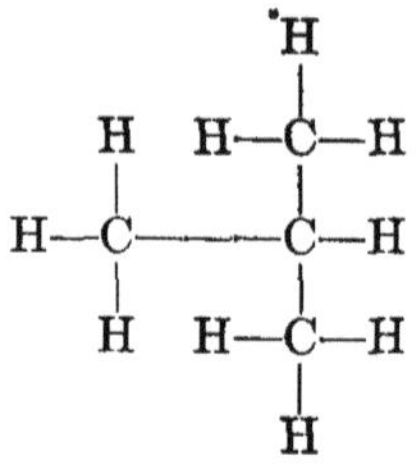

is 1046.4 kcal/mole.

These deviations are probably due primarily to a difference in the C—H bond energies since it has long been known in hydrocarbon chemistry that there is a difference in the reactivity of C—H bonds depending on the position in the chain. For example, it has been shown by E. W. R. Steacie [23] that the relative attack of alkyl radicals on primary, secondary, and tertiary hydrogen atoms is 1:3:4. Primary, secondary, and tertiary hydrogen atoms are hydrogen atoms attached to carbon atoms which are also bonded, respectively, to two, one, or no additional hydrogen atoms. However, these bonds can be treated as different bonds in order to allow for these deviations in numerical calculations.

Table B,16. After [24] by permission.

Molecule	C—C—C bond angle, degrees	C—C bond energy, kcal
Straight chain hydrocarbon	111 (Normal tetrahedral angle)	62.8
Cyclopropane	60	50.8
Cyclobutane	90	51.8
Cyclopentane	108	60.0
Cyclohexane	111	61.0

Another significant deviation from single bond constancy is exhibited by so-called "strained molecules," i.e. molecules for which the usual valence bond angles are distorted. These deviations are illustrated in Table B,16 and are due to the fact that the bonding orbitals do not overlap to the maximum extent.

Next, one must also consider multiple bonds.

Isolated double and triple bonds are bonds consisting of four and six electrons, respectively, overlapping entirely between two adjacent atoms. The average energies of these bonds are approximately constant as long as they are localized between two atoms; however, the energy in multiple bonds is not simply a multiple of the energy in the analogous single bonds, because the additional electrons are not in the same types of orbitals as those forming single bonds. For example, one finds a normal C—C single bond in ethane

$$
\begin{array}{c}
\text{H} \quad \text{H} \\
| \qquad | \\
\text{H—C—C—H} \\
| \qquad | \\
\text{H} \quad \text{H}
\end{array}
$$

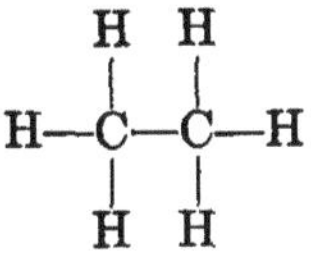

which is due to the overlap between two p electrons directed along the C—C bond—one from each carbon atom—and this bond has an average bond energy of 62.8 kcal or 79.1 kcal depending upon whether one uses $L_s = 124$ kcal/mole or $L_s = 170$ kcal/mole for the heat of sublimation of graphite.[17] The C—C double bond, such as occurs in ethylene

$$
\begin{array}{c}
\text{H} \quad \text{H} \\
| \qquad | \\
\text{C}{=}\text{C} \\
| \qquad | \\
\text{H} \quad \text{H}
\end{array}
$$

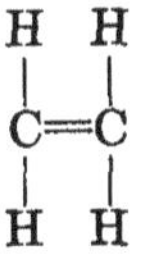

consists of the normal single bond and a second bond due to the overlap of two π electrons. These electrons are in approximately unexcited p orbitals with an axis normal to the single bond. Using the value $L_s = 124$ kcal/mole, the average bond energy of the C$=$C bond is 101.16 kcal/mole. The triple bond, as in acetylene

$$
\begin{array}{c}
\text{H} \quad \text{H} \\
| \qquad | \\
\text{C}{\equiv}\text{C}
\end{array}
$$

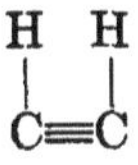

is 128.15 kcal/mole.

In molecules containing multiple bonds the average bond energy of the multiple bonds can be assumed constant if there is no ambiguity in the localization of the electrons between two atoms (isolated multiple bonds) and only one valence bond structure can be written. However, there are many molecules in which more than one equivalent valence bond structure can be written or in the language of molecular orbitals, there are certain "unsaturation" or "mobile electrons" which cannot be localized between any two adjacent atoms but rather move throughout the field of the entire molecule producing additional stabilization of the structure. This energy of stabilization has often been called resonance energy and is defined by the following relation:

$$
\sum \text{average bond energies} - \text{heat of atomization} = \text{resonance energy}
$$

[17] The determination of the heat of sublimation of graphite is experimentally difficult and authorities are sharply divided as to which of these two values is correct. However, even though the choice of this value affects the value of the bond energies it cancels out in prediction of heats of combustion which are more important in practical problems. For convenience, because this value has been more widely used in the literature we use the value $L_s = 124$ kcal/mole.

where the first term is obtained by assuming a certain valence bond structure and using numerical values for the average bond energies obtained from analogous molecules having localized multiple bonds. Benzene is the classical example for illustrating resonance.

Assuming a hypothetical structure for benzene, namely

and using experimentally determined values one finds for benzene

$$6(\text{C—H})\text{ bond energy} + 3(\text{C—C})\text{ bond energy} + 3(\text{C=C})\text{ bond energy} - \text{heat of atomization of benzene} \cong 34\text{ kcal/mole}$$

Thus resonance energy can be given a value if one knows the heat of atomization accurately. There have been many semiempirical and some theoretical calculations made to determine resonance energy and some progress has been made. However, the exact nature of these deviations from additivity in bond energy is much more significant in the understanding of molecular structure and chemical properties than in the assignment of values which enable one to predict heats of formation or heats of combustion.

B,17. Use of Average Bond Energies for Calculating Heats of Combustion. If the combustion of a molecule M can be represented by the following balanced chemical equation

$$n\text{M} + m\text{O}_2 \rightarrow a\text{P}_1 + b\text{P}_2 \qquad (17\text{-}1)$$

then the relationship between the heat of formation of a molecule and the heat of combustion of the molecule is given in the following equation

$$n\Delta H_f\text{ (molecule)} + mD = a\Delta H_f(\text{P}_1) + b\Delta H_f(\text{P}_2) + Q \qquad (17\text{-}2)$$

where $\Delta H_f(\text{M})$ = heat of formation of the molecule M in its gaseous state from the constituent atoms in their gaseous state, D = dissociation energy of O_2, $\Delta H_f[(\text{P}_1), (\text{P}_2)]$ = heats of formation of the products in their gaseous state, Q = heat of combustion of the molecule. However, the heats of formation of the products and the heats of combustion are usually measured in their standard state so that Eq. 17-2 has to be modified in order to use the usual experimental data. However, if one calculates the heat of formation of a molecule from the average bond energies one obtains the heat of formation of the molecule in its gaseous state. The calculation of the heat of combustion of a molecule from average bond energies and the necessary thermochemic data is now illustrated for the normal hydrocarbon C_nH_{2n+2} where the process of combustion can be represented by the following set of chemical equations:

$$
\begin{aligned}
&C_nH_{2n+2} + \frac{3n+1}{2} O_2 \rightarrow nCO_2 + (n+1)H_2O + Q(C_nH_{2n+2}) \\
&nCO_2 \rightarrow nC_{\text{solid}} + nO_2 - nQ(C) \\
&(n+1)H_2O \rightarrow (n+1)H_2 + \frac{n+1}{2} O_2 - nQ(H_2) \\
&(n+1)H_2 \rightarrow (2n+2)H_{\text{gas}} - D(H_2) \qquad (17\text{-}3) \\
&nC_{\text{solid}} + nL_s \rightarrow nC_{\text{gas}} \\
&nC_{\text{gas}} + (2n+2)H_{\text{gas}} \rightarrow C_nH_{2n+2} + \Delta H_f(C_nH_{2n+2})
\end{aligned}
$$

where $Q(M)$ = heat of formation of molecule M in its standard state
D = dissociation energy of molecular H_2
L_s = heat of sublimation of carbon
ΔH_f = heat of formation of C_nH_{2n+2}

From this cycle one can write the following thermochemical equation

$$\Delta H_f = nQ(C) + (n+1)Q(H_2) + nL_s + (n+1)D_{H_2} - QC_nH_{2n+2} \qquad (17\text{-}4)$$

Since the other quantities except L_s are known from accurate experimental data, one may calculate the heat of combustion $Q(C_nH_{2n+2})$ if the heat of formation ΔH_f (or vice versa) is known and one chooses[18] a value for L_s which is either 124.1 kcal/mole (Herzberg) or 170 kcal/mole (Gaydon and Penny). However, if neither ΔH_f nor $Q(C_nH_{2n+2})$ are known one can estimate $\Delta H_f(C_nH_{2n+2})$ according to the following equation

$$\Delta H_f(C_nH_{2n+2}) = (n-1)E_{C-C} + (2n+2)E_{C-H} \qquad (17\text{-}5)$$

using the values in Table B,17. In this case the calculated value of the heat of combustion is independent of L_s since L_s has been used to calculate the bond energies and thus cancels out if one uses the same value of L_s in Eq. 17-4 as is used in Table B,17. This method can be used for molecules other than hydrocarbons as long as the necessary experimental data are available and the molecules in question do not possess any appreciable resonance energy which cannot be estimated. Resonance energies for many characteristic structures have been experimentally determined and are discussed extensively in [*6,24*]. Another experimental value that is often important in such calculations and is also not settled is the dissociation energy of molecular N_2.

The validity of this method for n saturated hydrocarbons and n alco-

[18] See [*17*] and [*18*] for difficulties in determining L_s and for references to the now extensive literature relevant to the interpretation of these measurements.

hols has been established. As would be expected, the agreement is very good for the higher members of the series where the number of secondary hydrogen atoms is large.

The previous considerations have dealt largely with molecules made up of electron pair bonds for which the additivity assumption was approximately valid. However, there are a large number of molecules

Table B,17. Bond energies in organic molecules.*

Bond	*Energy, kcal*	*Bond*	*Energy, kcal*
C—H	85.56	C—F	104†
C—C	62.77	C—Cl	69
C=C	101.16	C—Br	57
C≡C (acetylene)	128.15	C—I	43
O—H	110.	N—H (NH_3)	83
C—O (alcohols)	75	N—C	53.5
C—O (ethers)	75	C≡N (HCN)	146
C=O (ketones)	155–157	C≡N (nitriles)	149
C=O (CH_2O)	144	N≡C (isonitriles)	139
C=O (aldehydes)	149.5	C≡N	84†
C(=O)—OH (HCOOH)	348	N=N	80†
C(=O)—OH (acids)	360†	N—N	27 ± 3
C(=O)—OC (formates)	313	N—O	61†
C(=O)—OC (esters)	327	N=O	108†
		N^{+}(=O)—O^{-}	$169 < E < 186$
		C=S (CS_2)	107.5
		C—S	54
		S—H	82

* Taken from [24, p. 240] by permission. This is based on the heat of sublimation of graphite = 124 kcal/mole.

† Approximate values.

which contain bonds called ionic bonds; these are formed in an entirely different way. In an ionic bond between AB the electron from atom B is transferred to the other atom A forming two ions A and B and these are then held together by the Coulombic attraction between the two charges of opposite sign. If this simple description were adequate it might be possible to devise some fairly simple scheme for determining bond energies and using these to calculate heats of formation of ionic compounds, but there are bonds which can be represented as a combination of ionic and covalent bonding in almost all proportions. However, this proportion depends very much on the other atoms present in the molecule. Accordingly, the whole scheme of bond energies is not very suitable for inorganic compounds. Thus, if one wishes to predict heats of formation or heats of combustion of organic molecules and is satisfied with an accuracy of ± 10 per cent, one can use average bond energies to estimate these quantities. As more experimental data are accumulated, and in particular, more bond dissociation energies are measured and better spectroscopic data are obtained, it should be possible to determine more accurately the heats of formation of molecules for which measurements do not exist. These also will contribute much to an understanding of both molecular structure and chemical reaction rate theory.

A recent volume, *A Discussion on Bond Energies and Bond Lengths* [*25*], gives an excellent up-to-date discussion of the various factors affecting bond strengths. Many more quantities than we have considered in this chapter are discussed. The authors are concerned with small differences in bond energies, 1 or 2 kilocalories. These factors have little effect in the prediction of heats of combustion as discussed here, but are exceedingly valuable in a better understanding of molecular structure.

CHAPTER 3. ACTIVATION ENERGIES

KARL F. HERZFELD
VIRGINIA GRIFFING

B,18. Introduction. The Arrhenius Equation. It has long been known that stable molecules react with one another to form new molecules; old bonds are broken, new bonds are formed, and atoms are rearranged. Furthermore, these reactions often take place with a finite measureable rate. If one reviews the experimentally determined rate data, accumulated in the past sixty years, two striking generalizations become evident: (1) the rate of a chemical reaction is strongly dependent on temperature, e.g. the rate of a reaction ordinarily doubles for each 10° rise in temperature, while the number of collisions between molecules varies scarcely at all (proportional to $T^{\frac{1}{2}}$) over a small temperature range; (2) the number of molecules reacting in unit time is usually much smaller than the number of collisions per unit time. These two facts suggest that only a restricted number of molecules present in an aggregate at a given time fulfill the requisite conditions for reaction.

Arrhenius was the first to formulate and explain these generalizations in terms of the following mechanism. He assumed that chemical reactions take place according to the following scheme:

$$\text{average molecules} \underset{k_2}{\overset{k_1}{\rightleftarrows}} \text{energetic molecules} \tag{18-1a}$$

$$\text{energetic molecules} \rightarrow \text{products of chemical reaction} \tag{18-1b}$$

If one makes the assumption that the forward reaction goes with a velocity constant k_1 and the back reaction k_2, and that the equilibrium between average molecules and "energetic molecules" is essentially maintained, the temperature variation of the equilibrium constant $K = k_1/k_2$ is given by the Van't Hoff equation

$$\frac{d \ln K}{dT} = \frac{q}{RT^2}$$

where q is the heat necessary to convert 1 mole of average molecules into energetic molecules. Then

$$\frac{d \ln k_1}{dT} - \frac{d \ln k_2}{dT} = \frac{q}{RT^2} \tag{18-2}$$

Now if

$$\frac{d \ln k_1}{dT} = \frac{E_{a1}}{RT^2} + B \tag{18-3a}$$

and E_{a2} is defined by $E_{a1} - E_{a2} = q$

$$\frac{d \ln k_2}{dT} = \frac{E_{a2}}{RT^2} + B \tag{18-3b}$$

If B is taken as zero then

$$\frac{d \ln k}{dT} = \frac{E_a}{RT^2} \tag{18-4}$$

or if E_a is temperature independent one obtains

$$k = Ce^{-\frac{E_a}{RT}} \tag{18-5}$$

which has the same form as the equation proposed empirically by Arrhenius as early as 1889. This equation can naively be interpreted as the product of the total number of collisions multiplied by the fraction of collisions resulting in chemical reaction. This interpretation is based upon the assumption that activation is due to collision between molecules. E_a is called the activation energy of the reaction and was considered by Arrhenius to be the energy that must be supplied to average molecules to change them into "energetic molecules" which could then react according to Eq. 18-1b. Even though the detailed notions of Arrhenius are erroneous, the fundamental idea with its later modifications is now almost universally accepted. It is now believed that any chemical process which goes at a finite rate requires an energy of activation which is the minimum energy the system must acquire before the process takes place. Examples of such processes which follow the Arrhenius law are

$$\frac{\text{number of molecules evaporating}}{\text{second}} = C_1 e^{-\frac{L}{RT}}$$

where L = molar latent heat, and

$$\frac{\text{number of electrons emitted by an incandescent solid}}{\text{second}} = C_2 e^{-\frac{W}{RT}}$$

where W = thermionic work function per electron.

Experimental rate data can be used to calculate the activation energy according to Eq. 18-3a if it can be shown experimentally that the constant B is zero. If $\ln k$ is plotted against $1/T$, a straight line with slope $-E_a/R$ is observed. Even though the constant C, usually called the "frequency factor" of the reaction, were proportional to T^n (where n is a small number) over the small range of temperatures for which it is practicable to measure k, the deviation from linearity is not significant. In fact, any marked deviation from linearity when $\ln k$ is plotted against $1/T$ can safely be attributed to complicated kinetics, such as consecutive reactions, chain reactions, etc. However, if the temperature dependence of the frequency factor is known, the experimental energy of activation, E_a, should be determined by plotting $\ln k/T^n$ against $1/T$.

B,19. Formal Theory of Reaction Kinetics. In order to give a more exact formulation of the concept of energy of activation it is convenient to classify chemical reactions according to the number of molecules taking part in each elementary step leading to chemical change. Since many elementary reactions take place according to a first order (unimolecular) or second order (bimolecular) mechanism, only these two mechanisms are considered. Most chemical reactions do not follow these simple kinetic laws but are due to a series of such elementary reactions taking place successively and simultaneously. For such reactions one can calculate an over-all energy of activation.

First order, unimolecular reactions. A reaction which follows a unimolecular mechanism can be described as follows: A normal molecule A is converted to an energetic molecule A′ which subsequently spontaneously decomposes or rearranges itself to form the product. A first order reaction is one for which the rate constant is defined by

$$-\frac{d[\text{A}]}{dt} = k_1[\text{A}] \tag{19-1}$$

Consider a first order, unimolecular reaction in which the rate of activation is sufficiently fast to maintain the equilibrium concentration of activated molecules. Then, according to the Maxwell-Boltzmann distribution law (see Art. 26) the number of molecules in the activated state, $N_{A'}$, is given by

$$N_{A'} = \frac{N_A p_A e^{-\frac{\epsilon_A}{kT}}}{\sum_i p_i e^{-\frac{\epsilon_i}{kT}}} \tag{19-2}$$

N_A = total number of molecules
p_A = a priori probability of the state A
ϵ_A = energy of the molecule in state A
Now assign to the molecules in state A the probability k_A of undergoing spontaneous reaction in unit time, the total rate of reaction is given by

$$-\frac{dN}{dt} = \frac{N\sum k_A p_A e^{-\frac{\epsilon_A}{kT}}}{\sum p_i e^{-\frac{\epsilon_i}{kT}}} \tag{19-3}$$

If the reaction is known to be first order, then

$$-\frac{dN}{dt} = k_1 N$$

and the specific rate is given by

$$k_1 = \frac{\sum k_A p_A e^{-\frac{\epsilon_A}{kT}}}{\sum p_i e^{-\frac{\epsilon_i}{kT}}} \tag{19-4}$$

and the temperature coefficient of the specific reaction rate is

$$\frac{d \ln k_1}{dT} = \frac{\sum k_A p_A e^{-\frac{\epsilon_A}{kT}} \frac{\epsilon_A}{kT^2} - \sum p_i e^{-\frac{\epsilon_i}{kT}} \frac{\epsilon_i}{kT^2}}{\sum p_i e^{-\frac{\epsilon_i}{kT^2}}}$$

or more simply

$$\frac{d \ln k_1}{dT} = \frac{\bar{\bar{\epsilon}} - \bar{\epsilon}}{kT^2} \tag{19-5}$$

and

$$\bar{\bar{\epsilon}} - \bar{\epsilon} = \text{energy of activation}$$

where $\bar{\bar{\epsilon}}$ = average energy of the molecules that react
$\bar{\epsilon}$ = average energy of all the molecules.
In most cases $\bar{\epsilon}$ will not differ appreciably from the average energy of the unactivated molecules, and thus $(\bar{\bar{\epsilon}} - \bar{\epsilon})$ will be the average energy that must be supplied to the molecules that react and is identical with the energy of activation proposed by Arrhenius.

Now if one examines the case where the rate of activation is not necessarily fast enough to maintain the equilibrium concentration of activated molecules, one obtains [*26*] for the temperature coefficient of the specific first order rate constant

$$\frac{d \ln k_1}{dT} = \overline{\frac{d \ln \theta_A}{dT}} + \overline{\frac{d \ln A_{A-i}}{dT}} + \frac{\bar{\epsilon}_A - \bar{\epsilon}_i}{kT^2} \tag{19-6}$$

where θ = fraction of activated molecules that return to the unactivated state

A_{A-i} = specific rate of deactivation back to any initial state i.

Since the process of deactivation can occur only by collisions or by the emission of radiation, the temperature coefficient of deactivation must be small; then for reactions of interest where the temperature coefficient is large, Eq. 19-6 becomes identical with Eq. 19-5 and $N_0(\bar{\epsilon} - \bar{\epsilon}) = E_A$, the energy of activation that occurs in the Arrhenius equation.

Second order, bimolecular reactions. A bimolecular mechanism can be described as follows:

$$A \rightarrow A' \quad \text{(activated molecule)}$$
$$B \rightarrow B' \quad \text{(activated molecule)}$$
$$A' + B' \rightarrow AB^*$$

or

$$A' + B \rightarrow AB^*$$
$$AB^* \rightarrow \text{products}$$

where AB*, called the reacting complex, represents the pair of molecules A′ and B′ at the instant that A′ is at the point of decomposing. B may or may not decompose. Such a reaction follows a second order law where the rate constant is given by

$$-\frac{d[A]}{dt} = k_2[A][B] \tag{19-7}$$

The species A may be the same as, or different from B.

Consider a second order bimolecular reaction in which it is assumed that the fraction of the two kinds of molecules A and B in the activated states A' and B' has the equilibrium value. This assumption will be valid only for reactions that remain second order over a range of concentrations. Then, using the Maxwell-Boltzmann equation, the number of activated molecules $N_{A'}$ and $N_{B'}$ is given by

$$N_{A'} = \frac{N_A p_{A'} e^{-\frac{\epsilon_{A'}}{kT}}}{\sum p_A e^{-\frac{\epsilon_A}{kT}}} \quad \text{and} \quad N_{B'} = \frac{N_B p_{B'} e^{-\frac{\epsilon_{B'}}{kT}}}{\sum p_B e^{-\frac{\epsilon_B}{kT}}} \tag{19-8}$$

where N_A and N_B are the number of molecules of A and B per cm^3, in the original state and p_A is the a priori probability of the state. Furthermore it can be shown from kinetic theory that the number of collisions between molecules depends on $T^{\frac{1}{2}}$; thus the rate of the bimolecular second order reaction is given by

$$-\frac{dN_A}{dt} = T^{\frac{1}{2}} \sum_{A',B'} k_{A'B'} N_{A'} N_{B'} \tag{19-9}$$

where $k_{A'B'}$ is a constant for each pair of activated molecules and is discussed in greater detail in Art. 20. Substituting Eq. 19-8 in Eq. 19-9 and solving for the second order rate constant

$$k_2 = -\frac{1}{N_A N_B}\frac{d(N_A)}{dt}$$

i.e.

$$k_2 = \frac{T^{\frac{1}{2}}\sum_{A'}\sum_{B'}k_{A'B'}p_{A'B'}e^{-\frac{\epsilon_{A'}+\epsilon_{B'}}{kT}}}{\left(\sum p_A e^{-\frac{\epsilon_A}{kT}}\right)\left(\sum p_B e^{-\frac{\epsilon_B}{kT}}\right)} \tag{19-10}$$

One can calculate the temperature coefficient and obtain

$$\frac{d \ln k_2}{dT} = \frac{1}{2T} + \frac{\overline{\overline{\epsilon_{A'} + \epsilon_{B'}}} - \overline{\epsilon_A + \epsilon_B}}{kT^2} \tag{19-11}$$

where $\overline{\overline{\epsilon_{A'} + \epsilon_{B'}}}$ = average energy of the molecules that react
$\overline{\epsilon_A + \epsilon_B}$ = average energy of all the molecules

Thus

$$N_0(\overline{\overline{\epsilon_{A'} + \epsilon_{B'}}} - \overline{\epsilon_A + \epsilon_B}) = E_a$$

the energy of activation in the Arrhenius equation.

The frequency factor for bimolecular reactions. The interpretation of the frequency factor, i.e. C, in the integrated form of the Arrhenius equation is more difficult. The process can be divided into two steps, the process of forming the activated complex AB*, and the process of decomposition of AB* into the products. Two extrême situations are as follows: In the first, the reaction AB* → products occurs almost immediately after the formation of AB*, before the complex can fly apart again into A′ + B, and before part of the energy of AB*, sufficient to make it inactive, can be removed by a further collision. In that case, all the AB* formed gives the products, and the rate of formation of the products is equal to the rate of formation of AB*. Accordingly, one can, in this case, interpret the frequency factor as the number of collisions between an A, and a B.

The other extreme occurs if most of the AB* complexes formed are destroyed before decomposing into the products after concentrating the necessary energy into a particular bond. The destruction might occur either by the flying apart of A′ and B or by the removal of sufficient energy through collision with another molecule. In this case, the rate at which the products are formed is determined by the rate of the decomposition of the complex AB*, and this rate may be calculated by the "absolute rate theory." The second case requires that the rate calculated by the absolute rate theory is *less* than would be given by the number of collisions forming the complex AB* (first extreme).

In the first case, the factor C can be interpreted in terms of the elementary collision theory of the kinetic theory of gases for a simple second order bimolecular reaction

$$A + B \rightarrow C + D$$

If one writes the rate in molecular units

$$v = kN_AN_B = ze^{-\frac{E_a}{RT}} \frac{\text{molecules}}{\text{cm}^3 \text{ sec}} \tag{19-12}$$

then z can be interpreted as the number of collisions per cm^3 per second and according to kinetic theory is

$$z = N_AN_B\sigma_{AB}^2 \left(\frac{8\pi kT}{\mu_{AB}}\right)^{\frac{1}{2}} \tag{19-13}$$

where N_A and N_B are the concentrations of reactants in number of molecules per cm^3, σ_{AB} is the average molecular diameter of A and B, and $\mu_{AB} = m_Am_B/(m_A + m_B)$, the reduced mass of the molecules A and B.

Thus

$$k = \sigma_{AB}^2 \left(\frac{8\pi kT}{\mu_{AB}}\right)^{\frac{1}{2}} e^{-\frac{E_a}{RT}} \tag{19-14}$$

Since σ_{AB} can be determined from viscosity data and E_a can be determined from rate data, the validity can be tested. It does, in fact, give good agreement for many simple reactions in both the gaseous and liquid states as shown by the values given in Tables B,19a and B,19b.

Table B,19a. Gaseous reactions.*

Reaction	E_a, calories	T^0, abs	$\sigma_{AB} \times 10^8$, cm	k_{obs}, liters/g molecule sec	k_{calc}, liters/g molecule sec	$\frac{k_{obs}}{k_{calc}}$
$2N_2O \rightarrow 2N_2 + O_2$	58,500	1001	3.3	3.80×10^{-1}	3.15×10^{-2}	12
$CH_3CHO \rightarrow CH_4 + CO$	45,500	800	5.0	5.15×10^{-1}	1.38×10^{-1}	3.7
$2HI \rightarrow H_2 + I_2$	43,800	666	3.5	2.20×10^{-4}	4.68×10^{-4}	0.47
$C_2H_4 + H_2 \rightarrow C_2H_6$	42,400	787	2.0	1.77×10^{-2}	3.53×10^{-1}	0.05
$H_2 + I_2 \rightarrow 2HI$	38,800	660	2.0	1.63×10^{-2}	3.46×10^{-2}	0.47
$2NO_2 \rightarrow 2NO + O_2$	26,500	610	3.3	2.06×10^{0}	6.93×10^{1}	0.03

* Table taken by permission from Moelwyn-Hughes, E. A., *The Kinetics of Reactions in Solution*, p. 74. Clarendon Press, Oxford, 1933.

However, there are many bimolecular reactions whose experimental rates deviate widely from the rate calculated according to the simple collision theory; for some reactions the calculated rates are larger and for others smaller than the observed rate. In order to account for these low values it has been observed that the number of collisions effective for reaction may be considerably smaller than the total number of collisions, since some critical orientation of even energetic molecules before the collision would be required to bring about reaction. In order to take this effect into account, Eq. 19-12 is modified thus

$$v = Pze^{-\frac{E_a}{RT}} \tag{19-15}$$

where P is called the steric factor and is the probability that a sufficiently energetic collision would be geometrically oriented to bring about reaction. This factor must be estimated for individual reactions and there is no a priori method for determining it.

Table B,19b. Liquid reactions.[*]

Reaction	Solvent	E_a, calories	$z \times 10^{-11}$ Observed	$z \times 10^{-11}$ Calculated	$\frac{k_{calc}}{k_{obs}}$
$CH_3ONa + 1:2:4$-$ClC_6H_2(NO_2)_2$	CH_3OH	17,450	1.91	2.42	1.3
$C_2H_5ONa + CH_3I$	C_2H_5OH	19,490	2.42	1.93	0.8
$C_2H_5ONa + C_2H_5I$	C_2H_5OH	20,650	1.49	2.23	1.5
$C_6H_5CH_2ONa + C_4H_9I$	C_2H_5OH	21,560	2.92	2.43	0.8
$C_6H_5CH_2ONa + iso\ C_4H_9I$	C_2H_5OH	21,350	2.45	2.43	1.0
$C_6H_5CH_2ONa + C_{16}H_{33}I$	C_2H_5OH	21,090	1.26	3.12	2.5
β-$C_{10}H_7ONa + C_2H_5I$	C_2H_5OH	19,840	0.11	2.21	20.1
$(CH_3)_2SO_4 + NaCNS$	CH_3OH	17,360	0.19	1.91	10.0
$NH_4CNO \rightarrow (NH_2)_2CO$	H_2O	23,170	42.7	4.05	0.1
$C_6H_4\overset{CH_2}{\underset{CO}{\diagup\ \diagdown}}O + OH^-$	H_2O	12,500	41.7	2.93	0.07
$C_6H_5CO[CH_2]_2Cl + I^-$	$(CH_3)_2CO$	22,160	10.5	1.88	0.2
$CH_3S[CH_2]_2Cl + I^-$	$(CH_3)_2CO$	20,740	0.085	1 57	18.5
$C_2H_5Br + OH^-$	C_2H_5OH	21,400	4.30	3.86	0.9
$C_2H_4Br_2 + I^-$	CH_3OH	25,100	1.07	1.39	1.3

[*] Table taken by permission from Moelwyn-Hughes, E. A., *The Kinetics of Reactions in Solution,* p. 79. Clarendon Press, Oxford, 1933.

However, it is even more probable that in the case where C is found too small, the second case is present and one has to use the absolute rate theory. In case of values of C large compared to $z/N_A N_B$ one must consider, with Lindemann and Polanyi, that more than one degree of freedom may be involved as the source of the activation energy. If the activation energy E_a may be distributed over s classical degrees of freedom in molecule A, then

$$\frac{N_{A'}}{N_A} = \left(\frac{E_A}{kT}\right)^{s-1} \frac{1}{(s-1)!} e^{-\frac{E}{kT}} \tag{19-16}$$

and Eq. 19-13 should be replaced by

$$z = N_A N_B \sigma_{AB}^2 \left(\frac{8\pi kT}{\mu_{AB}}\right)^{\frac{1}{2}} \left(\frac{E_A}{kT}\right)^{s-1} \frac{1}{(s-1)!} \tag{19-17}$$

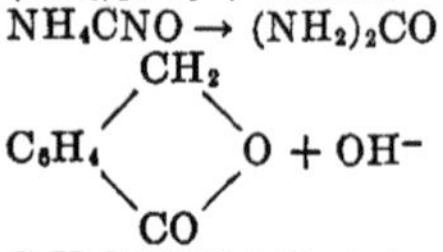

If, e.g. $E_A/kT = 20$, $s = 5$, the increase in z is by a factor of more than 6,000.

The difficulty in interpretation lies in the fact that in general, s is unknown. If s is sufficiently large, the rate of supply of activated complexes may be so large that the decomposition of the complex into the products (absolute rate theory) may be determining, even if that rate is larger than the simple collision theory, with $s = 1$, would give.

The idea of entropy of activation, associated with the steric factor, was developed by Rodebush, La Mer, and others. However, this concept is more precise from the point of view of the absolute reaction rate theory discussed in Art. 20.

Frequency factor in unimolecular reactions. Here also, the energy supply must occur through collisions which are bimolecular. To give a resultant monomolecular reaction, the decomposition of the activated complex must be slow compared to the energy supply. Originally, this provided a difficulty, since according to the forerunners of the absolute rate theory (Herzfeld), and according to experiment, C is of the order 10^{12} to 10^{13} while the simple collision factor under standard conditions is 10^9. Actually, in diatomic molecules decomposition is not unimolecular, but bimolecular in consequence of the recombination being termolecular. For polyatomic molecules, the solution lies, as Lindemann and Polanyi have shown, in an increase of the collision factor according to Eq. 19-17.

In Lindemann's picture the energy of activation is accumulated in the internal degrees of freedom of the molecule before reaction can occur. Since this energy is distributed among all the vibrational degrees of freedom and must get into one bond before reaction can occur (i.e. probably a particular linear combination of the normal modes of the molecule) there would be a time lag between activation and reaction which would be proportional to the reciprocal of the frequency factor. Unimolecular reactions become second order at low pressures for which the rate of deactivation due to collisions is much smaller than the rate of decomposition.

For termolecular reactions, i.e. a reaction which requires the simultaneous collision of three particles, the energy of activation must be very small or the reaction would not go at all since the probability of the occurrence of a triple collision is so small. Only a few reactions following third order kinetics have been observed and in all these cases the energy of activation is small (see [*27*]). However, ternary collisions are important in the recombination of atoms and seem to be a relatively efficient process. It is now a well-established fact that in the recombination of atoms, a third body is required to carry off the excess energy or the two atoms would again separate. Since many atoms do recombine with a measureable rate, one can conclude that the recombination of atoms does not require an energy of activation.

B,20. Theory of Absolute Reaction Rates. More recently, Eyring and Polanyi [*28*], and others have given the following general picture of the process of decomposition of the activated complex. Assume a reaction goes according to the relationship

$$\text{reactants} \rightarrow \text{products}$$

with a finite rate. Then if one plots the potential energy as a function of all the coordinates of the constituent atoms one would obtain a potential energy surface which would contain at least two minima. However, since the reaction goes at a finite rate these two minima must be

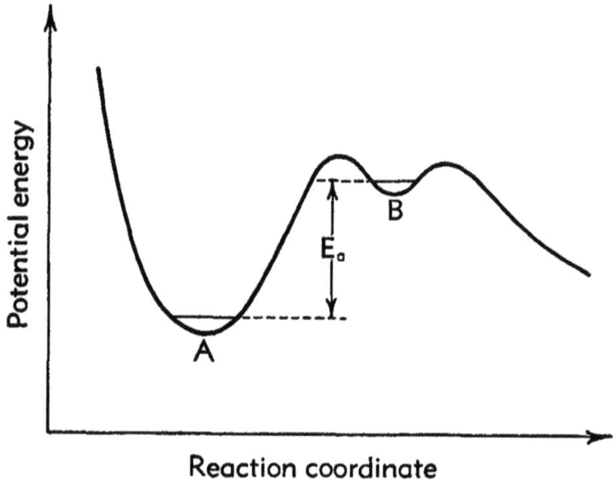

Fig. B,20. A = stable molecule; B = activated complex; E_a = activation energy (for the formation of the activated complex); i.e. difference between the zero point energies of the stable configuration A and the activated complex B.

separated by potential barriers. If one connects the point representing the stable configuration of the reactants with that representing the stable products by a continuous curve on the potential energy surface choosing as the reaction coordinate the path that surmounts the minimum barrier, one obtains the one-dimensional potential energy curve given in Fig. B,20.

According to this picture, the energy of activation is the minimum energy which must be supplied to the reactant molecules at 0°K to raise them to the point B, i.e. zero point energy of the activated complex. The molecular complex at this point is called the "activated complex." Then, using this concept, the reaction rate is the number of activated complexes passing over the potential energy barrier per second. This rate is equal to the concentration of the activated complex multiplied by the average speed with which the complex moves. In order to calculate the concentration of activated complexes for a simple bimolecular reaction, consider the reaction

$$A + B \rightleftarrows (AB)^{\ddagger} \rightarrow \text{products}$$

Assume $(AB)^{\ddagger}$ in equilibrium with the reactants; then the equilibrium constant for the formation of $(AB)^{\ddagger}$ is given by

$$K^{\ddagger} = \frac{[(AB)^{\ddagger}]}{[A][B]}$$

where [A] means concentration of A, etc.; then

$$\frac{d[A]}{dt} = K^{\ddagger}[A][B] \times \text{rate of passage over the barrier}$$

According to this theory, one then assumes that the complex spontaneously decomposes into its products when one of its vibrations has its classical energy $\epsilon = kT$ and becomes a translation, or the rate of passage over the barrier becomes kT/h. Then the reaction rate becomes

$$\frac{d[A]}{dt} = K^{\ddagger}[A][B] \times \frac{kT}{h} \tag{20-1}$$

(For a unimolecular reaction, $K^{\ddagger}$ would be the equilibrium constant for the equilibrium $A \rightarrow A^{\ddagger}$.) According to statistical mechanics, the equilibrium constant $K^{\ddagger}$ is

$$K^{\ddagger} = \frac{Q^{\ddagger}}{Q_A Q_B} e^{-\frac{E_a}{RT}} \tag{20-2}$$

where $Q^{\ddagger}$, Q_A, and Q_B are the partition functions of the activated complex and of the molecules A and B respectively (or $A^{\ddagger}$ in unimolecular reactions) and E_a is the difference in the zero point energy of the activated complex and the reactants, i.e. the energy of activation shown in Fig. B,20. In $Q^{\ddagger}$ the vibration which has become a translation is to be omitted.

$$k_2 = \frac{kT}{h} \frac{Q^{\ddagger}}{Q_A Q_B} e^{-\frac{E_a}{RT}} \tag{20-3}$$

Thermodynamically

$$\Delta F = -RT \ln K^{\ddagger}$$

and

$$\Delta F^{\ddagger} = \Delta H^{\ddagger} - T\Delta S^{\ddagger}$$

and Eq. 20-3 becomes

$$k_2 = \frac{kT}{h} e^{+\frac{\Delta S^{\ddagger}}{R}} e^{-\frac{\Delta H^{\ddagger}}{RT}} \tag{20-4}$$

According to the definition of the experimental activation energy calculated according to the Arrhenius equation (i.e. Eq. 18-5)

$$\frac{d \ln k}{dT} = \frac{E_a}{RT^2} = \frac{\Delta H^{\ddagger} + (\Delta n^{\ddagger} - 1)RT}{RT^2}$$

where $\Delta n^{\ddagger}$ = the number of moles of "activated complex" − number of moles of reactants. Thus the "entropy of activation" can be calculated from the experimental rate constant and activation energy.

In comparing Eq. 20-3 with Eq. 19-10 one has

$$\frac{kT}{h}\frac{Q^{\ddagger}}{Q_{A}Q_{B}} = \frac{T^{\frac{1}{2}}\sum\sum k'_{AB}p_{A'B'}e^{\frac{\frac{1}{N_0}E_a-\epsilon_{A'}-\epsilon_{B'}}{kT}}}{\left(\sum p_{A}e^{-\frac{\epsilon_A}{kT}}\right)\left(\sum p_{B}e^{-\frac{\epsilon_B}{kT}}\right)}$$

From Eq. 20-4, the frequency factor, according to this theory, is

$$C = \frac{kT}{h}\cdot\frac{\text{product of the partition functions for the activated complex}}{\text{product of the partition functions of the reactant molecules}}$$

B,21. Theoretical Calculation of the Structure of the Activated Complex. The activated complex can be treated as a definite molecular species. Thus the structure of the activated complex can be calculated, in principle, according to the methods of quantum mechanics. Just as in other problems of molecular structure, one can write the Schrödinger equation as a function of the coordinates of all the electrons and nuclei of the reactants. Then, this wave equation must be solved as a function of all the coordinates, without any assumptions of electron pairs or minimum energy. The energy of the activated state would then be the highest energy through which the system must pass in going from the initial configuration (reactants) to the final configuration (products). The results of such a calculation would be a many-dimensional potential energy surface. In practice, such a calculation is not feasible for even the simplest molecules and reactions. Wigner *[29]* and Evans and Polanyi *[30]* have discussed the inherent difficulties of calculating such potential energy surfaces by anything approaching rigorous methods.

B,22. Semiempirical Calculations on Activation Energies.[19] Eyring and his collaborators have used a semiempirical method for calculating the structure of the activated complex for a simple reaction

$$AB + C \rightarrow BC + A$$

where A, B, and C are all monatomic. One can show that the atom C will approach the molecule AB with the minimum energy along a line formed by the extension of a straight line connecting the atoms A and B. Call the distance between A and B, r_{AB} and the distance between B and C, r_{BC}. Then when r_{BC} is large, r_{AB} will be the interatomic distance of the stable molecule AB; similarly, after reaction, r_{BC} will be the interatomic

[19] This method is discussed in detail in *[31]*. Literature reference for applications to individual reactions is also found in this book.

distance of BC and r_{AB} will be large. If one plots the energy as a function of these two coordinates, one will obtain a potential energy surface along which the chemical reaction must take place. The reaction coordinate will be a continuous path along which a particle can move with a minimum expenditure of energy from the stable well representing reactants to the stable configuration of the products. The height of the potential hill between the two low points will be the energy of activation. Eyring and his collaborators have empirically calculated potential energy surfaces for a few reactions. It is assumed that all directional valence effects can be neglected; a Morse (see Art. 15) potential energy curve for each pair of atoms is constructed (i.e. energy as a function of distance). This energy can be divided into two parts, the electrostatic Coulomb energy and a quantum mechanical form of energy called exchange energy. Since these energies depend on the coordinates differently, Eyring quite arbitrarily assumes a ratio of 14/86 for Coulomb energy/exchange energy. This value was chosen to give agreement between experimental and calculated activation energies for ortho-para hydrogen conversion. He then uses this same ratio in considering all nonmetallic electron pair bonds. The quantitative agreement with experiment which Eyring obtains is surprisingly good considering the radical assumptions that have been made. In fact, the agreement is destroyed when less drastic assumptions are made. Slater [*3*] suggests that the agreement may be due to the fact that it is really a method-of-interpolation formula between the energies of the final and initial states. Furthermore, direct computations show that the assumed energy ratio between Coulombic and exchange energy varies considerably from atom to atom (for p electrons the exchange energy sometimes becomes negative). However, Eyring's calculations have been useful in giving a general concept of the activated complex as a problem of molecular structure and of the energy of activation as the difference in the binding energy of this complex and that of the stable reactants.

B,23. Discussion. In general, one can conclude that reactions which require a high energy of activation will proceed slowly. In practice, the only dependable method at present of determining the energy of activation of a reaction is to measure its rate as a function of temperature, to attempt to estimate the temperature dependence of the frequency factor, and to calculate the energy of activation from the experimental rate data. If one can calculate the energy of activation in this way, one can then get an independent check by calculating, e.g. the collision cross section σ_{AB} (collision theory) and comparing this with an experimental value of σ_{AB} from viscosity data. However, certain general statements can be made. For endothermic reactions, the activation energy is of the same order of magnitude as the heat of reaction plus the energy of activation of the reverse process. Szwarc has used this conclusion and makes the assumption that the recombination of radicals does not require an energy of activation in determining the bond dissociation energy of a molecule into free radicals (usually defined as electrically neutral fragments of a molecule with an odd number of electrons). His values of bond dissociation energies seem to be in agreement with those determined by other methods [*20*]. For exothermic reactions of the type

$$AB + C \rightarrow BC + A$$

where AB is a diatomic molecule and C is a single atom, the energy of activation is almost zero, e.g.

$$Na + Cl_2 \rightarrow NaCl + Cl$$

The recombination of atoms does not require an appreciable energy of activation. For exothermic reactions in which diatomic molecules exchange partners, i.e.

$$AB + CD \rightarrow AC + BD + \text{heat}$$

it has been found that the activation energy is approximately one-fourth the sum of the dissociation energies of the reactant molecules, e.g.

$$\begin{aligned} H_2 + I_2 \rightarrow 2HI \qquad E_a\,(\text{exptl}) &= 40 \text{ kcal/mole} \\ \tfrac{1}{4}(D_{H_2} + D_{I_2}) &= 40 \text{ kcal/mole} \\ H_2 + H_2 \rightarrow H_2 + H_2 \qquad E_a\,(\text{exptl}) &> 57 \text{ kcal/mole} \\ \tfrac{1}{2}(D_{H_2}) &= 52 \text{ kcal/mole} \end{aligned}$$

One would expect that often the activation energy would be less than the energy required to break bonds since new bonds are formed in the formation of the activated complex while old bonds are only weakened. In semiempirical calculations [*31*] it has been found that a bond length increases in the activated state about 10 per cent of its normal value.

CHAPTER 4. STATISTICAL MECHANICS[20]

J. O. HIRSCHFELDER C. F. CURTISS
R. B. BIRD E. L. SPOTZ

The methods of classical and quantum mechanics have been highly successful in the analysis of the behavior of simple systems involving only a few degrees of freedom. It is clearly impractical, however, to apply these principles to the examination of the behavior of a very

[20] This chapter is based upon material given in Chapter 2 of *Molecular Theory of Gases and Liquids*, by J. O. Hirschfelder, C. F. Curtiss, and R. B. Bird (John Wiley & Sons, Inc., 1954). The authors wish to thank the publishers for permission to use this material.

complex system, such as a gas containing 6.023×10^{23} molecules. Furthermore a knowledge of the microscopic behavior thus provided is not particularly of interest. When the knowledge of the initial state of the system is limited to the results of measurements of the bulk properties, statistical mechanics may be used to predict the change of the macroscopic state of the system with time.

B,24. The Fundamentals of Statistical Thermodynamics. Statistical mechanics may be thought of as being made up of two important branches: *equilibrium* and *nonequilibrium* statistical mechanics. The former is well understood from the standpoint of the fundamental principles and the formal development of the theory. The important results of the theory are summarized in this chapter. Although progress in this direction has thus far been seriously handicapped by the numerical difficulties encountered, much has been learned about the properties of matter and their dependence on the forces between the molecules.

The nonequilibrium statistical mechanics is of more recent development and is one of the frontier fields of research today. A general discussion of the formal treatment of nonequilibrium statistical mechanics is presented in Chap. 5 and the applications to hydrodynamics and transport properties are discussed in Sec. D.

In this article we discuss the statistical mechanical relations between the thermodynamic quantities (such as the internal energy, entropy, and temperature) and the properties of molecules. The relationship between statistical mechanics and thermodynamics is usually discussed in terms of the "partition function." Hence we preface the treatment of statistical thermodynamics by a brief discussion of the partition function in quantum and classical statistics.

The partition function. The quantum mechanical partition function[21] (which is sometimes referred to as the "sum over states" or "Zustandssumme") for a system of N molecules, Q_{N_q}, is defined by either of the two equivalent relations:

$$Q_{N_q} = \underset{\substack{\text{Sum over all}\\ \text{energy } \textit{states}}}{\sum e^{-\beta E_j}} = \underset{\substack{\text{Sum over all}\\ \text{energy } \textit{levels}}}{\sum g_j e^{-\beta E_j}} \tag{24-1}$$

In the first expression E_j represents the energy of the system in the jth quantum state. On the other hand, in the second expression E_j is the energy associated with the jth energy level, and the degeneracy of the level is indicated by g_j (that is, the jth level is composed of g_j states). Both forms for the partition function are in common use in the literature.

[21] Both the classical and quantum mechanical partition functions are designated by Q_N. When it is desired to distinguish between the two quantities, a subscript q serves to indicate the quantum mechanical partition function.

The partition function is a function of β (which is shown presently to be $1/kT$) and also of any mechanical properties of the system which influence the energy levels. Usually the only mechanical parameter of the system which enters is the volume. In Art. 25 the explicit nature of the volume dependence of the E_j is discussed in connection with practical thermodynamical calculations.

The quantum mechanical partition function may also be written as an integral over configuration space, involving the quantum mechanical Hamiltonian operator, $\mathcal{H}$:

$$Q_{N_q} = \int \sum_\rho \phi_\rho^*(\mathbf{r}^N) e^{-\beta\mathcal{H}} \phi_\rho(\mathbf{r}^N) d\mathbf{r}^N \tag{24-2}$$

In this expression the $\phi_\rho(\mathbf{r}^N)$ form a complete set of orthonormal functions with quantum numbers $\{\rho\} \equiv \rho_1, \rho_2, \rho_3 \cdots$. The symbol $\mathbf{r}^N \equiv \mathbf{r}_1, \mathbf{r}_2, \mathbf{r}_3, \cdots$ is a vector in $3N$-dimensional configuration space which gives the positions of the N particles in the system. When the partition function is written in this fashion, the volume dependence manifests itself in the limits of the integration over configuration space, and also in the boundary conditions which the expansion functions satisfy.

In the correspondence limit, where classical behavior is approached, it may be shown[22] that the partition function given in Eq. 24-2 becomes an integral over the classical phase space involving the classical Hamiltonian $H(\mathbf{r}^N, \mathbf{p}^N)$:

$$Q_N = [N!h^{3N}]^{-1} \iint e^{-\beta H(\mathbf{r}^N, \mathbf{p}^N)} d\mathbf{r}^N d\mathbf{p}^N \tag{24-3a}$$

or in terms of the potential energy of the system $\Phi(\mathbf{r}^N)$,

$$Q_N = \frac{(2\pi mkT)^{3N/2}}{N!h^{3N}} \int e^{-\beta\Phi(\mathbf{r}^N)} d\mathbf{r}^N \tag{24-3b}$$

These expressions for the classical partition function[23] are valid only for the case of a system of N identical particles. The factor $N!$, which appears in the denominator, is due to the Pauli exclusion principle. Because of the identity of the particles certain regions of phase space are equivalent in that they correspond to a simple renumbering of the particles. Since there are $N!$ permutations of the set of N particles, the factor $(1/N!)$ "corrects" for this equivalence. Such corrections must always be made in classical formulas. In the remainder of this article the quantum mechanical form of the partition function is used.

The internal energy and the first law of thermodynamics. Let us consider a large number of identical systems in weak energy contact with one another. This "ensemble" of systems represents a single system

[22] See for example the very excellent review article by J. de Boer [32].

[23] The classical partition function is sometimes called the "phase integral." This terminology should not be confused with the Sommerfeld-Wilson phase integrals of old quantum theory.

bathed in a thermostat at a particular temperature. It may be shown that the probable number of systems in such an ensemble which are in state j is proportional to $e^{-\beta E_j}$. Hence the average energy of the members of the ensemble which is the thermodynamic internal energy U is

$$U = \frac{\sum E_j e^{-\beta E_j}}{\sum e^{-\beta E_j}} \tag{24-4}$$

The internal energy may also be expressed directly as a derivative of the natural logarithm of the partition function:

$$U = -\left(\frac{\partial \ln Q_N}{\partial \beta}\right)_V \tag{24-5}$$

where the subscript V indicates that the external mechanical parameters (such as the volume) of the system are held fixed.

The first law of thermodynamics may be written in the differential form

$$dU = \delta q - \delta w \tag{24-6}$$

in which δq is the small quantity of heat absorbed by the system and δw is the work done by the system while undergoing a small change in state. Here, dU is an *exact* differential, while δq and δw are both *inexact* differentials. It is also possible to write an expression for dU in terms of the energy levels of the system. Letting $\bar{a}_j = Q_N^{-1} e^{-\beta E_j}$ we may write Eq. 24-4 as

$$U = \sum \bar{a}_j E_j \tag{24-7}$$

whence

$$dU = \sum E_j d\bar{a}_j + \sum \bar{a}_j dE_j \tag{24-8}$$

The first term on the right-hand side represents the change in energy due to a redistribution of the total energy over the various quantum states of the system. The second term gives the change in energy which results from the shift in the energy levels of the system caused by the alteration of the volume (or other external parameters). This latter term may clearly be associated with δw and the former with δq:

$$\delta q = \sum E_j d\bar{a}_j \tag{24-9}$$

$$\delta w = -\sum \bar{a}_j dE_j \tag{24-10}$$

Eq. 24-8 may then be regarded as giving a microscopic interpretation of the first law of thermodynamics.

Temperature and entropy and the second law of thermodynamics. It is now necessary to introduce the concepts of temperature and entropy. In the axiomatic thermodynamics the reciprocal of temperature is defined as the integrating factor of the heat change, and entropy is defined in terms of the perfect differential so obtained. The statistical definition of entropy is based on an analogy with this approach. We have obtained expressions which are intuitively related to the concepts of infinitesimal heat and work terms. The reciprocal of the temperature is defined as the integrating factor for this infinitesimal heat change and the entropy is defined by the perfect differential which results.

Let us define a function, the "entropy," by the relation

$$S = S(E_1, E_2, \cdots, \beta) = k(\ln Q_N + \beta U) + S_0 \qquad \text{(24-11)}$$

in which k is Boltzmann's constant and S_0 is a constant. It is now possible to define a perfect differential dS as

$$\begin{aligned} dS &= \left(\frac{\partial S}{\partial \beta}\right)_{E_1E_2\cdots} d\beta + \sum \left(\frac{\partial S}{\partial E_j}\right)_\beta dE_j \\ &= kd(\ln Q_N + \beta U) \end{aligned} \qquad \text{(24-12)}$$

Using the definition of the partition function and the relations in Eq. 24-9 and 24-10 for δq and δw, we get:

$$\begin{aligned} dS &= -\frac{k}{Q_N}\sum (\beta dE_j + E_j d\beta)e^{-\beta E_j} + k\beta dU + kUd\beta \\ &= k\beta\left[dU - \sum \bar{a}_j dE_j\right] \\ &= k\beta\delta q \equiv \frac{1}{T}\,\delta q \end{aligned} \qquad \text{(24-13)}$$

This demonstrates that $k\beta$, or $1/T$, is an integrating factor for the heat change. That is, multiplication of the inexact differential δq by $1/T$ yields an exact differential dS.

This in itself does not define the temperature T uniquely. However, if in addition it be required that the entropy be an extensive property then it can be shown that the choice is unique except for a multiplicative constant (the scale factor). An empirical temperature is often defined by the perfect gas thermometer. This temperature is related to the equation of state of a perfect gas by the equation $pV = NkT$. The thermodynamic temperature just defined is identical with this temperature since, as may easily be shown, the present treatment leads to the same equation of state. We may hence rewrite the expression for the entropy as

$$S = k\left[\ln Q_N + \frac{U}{kT}\right] + S_0 \qquad \text{(24-14)}$$

In this way it is possible to introduce temperature and entropy by a method analogous to thermodynamics. Clearly S and T are both state functions.

To complete the statistical proof of the second law of thermodynamics it would be necessary to show that the entropy of an isolated system never decreases. This is the content of the famous H-theorem, the proof of which is much too lengthy to be included here.[24]

Entropy at absolute zero and the third law of thermodynamics. The constant S_0, which occurs in the definition of entropy just given, is independent of the temperature and the mechanical properties of the system. In the limit as the temperature goes to absolute zero

$$\begin{aligned}\lim_{T\to 0} S &= \lim_{T\to 0} k\left[\ln \sum g_j e^{-E_j/kT} + \frac{\partial \ln \sum g_j e^{-E_j/kT}}{\partial \ln T}\right] + S_0 \\ &= k[\ln g_0] + S_0 \end{aligned} \tag{24-15}$$

in which g_0 is the degeneracy of the ground state of the system.

Let us now consider two states of the same system: One state consists of the chemical AB in a vessel at absolute zero; the other state consists of two vessels—one of A and one of B—both at absolute zero. It is possible to conceive of a reversible process by which the system may be transferred from one state to the other. Hence the two states AB and $A + B$ are really two states of the same system, differing only in the mechanical parameters describing the state. Since S_0 is the same for both states, the entropy difference in the states of the system is

$$S_{AB} - S_{A+B} = k \ln (g_{AB}/g_A g_B) \tag{24-16}$$

If all three materials, A, B, and AB are such that their ground states are nondegenerate, then $g_A = g_B = g_{AB} = 1$ and

$$\Delta S = S_{AB} - (S_A + S_B) = 0 \tag{24-17}$$

For many systems the lowest state is normally considered to be multiply degenerate. However, there is usually a small separation of the energy levels due to small perturbations which are ordinarily neglected. As kT approaches zero, these very small separations become effectively large compared with kT. Hence it is probable that, strictly speaking, the ground state is always nondegenerate and Eq. 24-17 is valid.

Inasmuch as S_0 always cancels out, it is convenient to let it be exactly zero and define the entropy as

$$S = k \ln Q_N + \frac{U}{T} \tag{24-18}$$

With this definition, the third law of thermodynamics assumes the form

$$\lim_{T\to 0} S = k \ln g_0 \tag{24-19}$$

and if the ground state of the system is nondegenerate

$$\lim_{T\to 0} S = 0 \tag{24-20}$$

[24] See for example R. H. Tolman [*33*] for a summary of the proofs of this theorem in classical and quantum statistics.

The quantity g_0 is the true degeneracy of the lowest quantum state of the system. However, in making comparisons with calorimetrically measured entropies it may be necessary to modify somewhat the meaning of this quantity. For if the separation of the lowest levels of the system is small compared with kT_0, where T_0 is the lowest measurable temperature, there would be a small hump in the specific heat curve at a lower temperature. The contribution of this hump to the entropy would be ignored in the integration of experimental C_p/T data. For this purpose it is necessary to consider as degenerate the group of levels separated by energies small compared to kT_0, where T_0 is the lowest temperature attained in the experimental measurements.

The thermodynamic properties in terms of the partition function. We have thus seen how the fundamental thermodynamic functions, energy and entropy, can be expressed in terms of the partition function which in turn depends upon the energy levels of the system. Therefore, if detailed information about the energy levels of a system is available, it is possible to calculate all of the thermodynamic properties from the partition function. The relations needed for this may be summarized as follows:

$$U = kT^2\left(\frac{\partial \ln Q_N}{\partial T}\right) \tag{24-21}$$

$$C_V = \frac{\partial}{\partial T}\left(kT^2\frac{\partial \ln Q_N}{\partial T}\right) \tag{24-22}$$

$$S = k \ln Q_N + \frac{U}{T} \tag{24-23}$$

$$A = U - TS = -kT \ln Q_N \tag{24-24}$$

For systems in which the only external mechanical parameter is the volume V, we have the additional relations

$$p = -\left(\frac{\partial A}{\partial V}\right)_T = kT\left(\frac{\partial \ln Q_N}{\partial V}\right)_T \tag{24-25}$$

$$H = U + pV = kT^2\left(\frac{\partial \ln Q_N}{\partial T}\right)_V + kTV\left(\frac{\partial \ln Q_N}{\partial V}\right)_T \tag{24-26}$$

$$F = H - TS = -kT \ln Q_N + kTV\left(\frac{\partial \ln Q_N}{\partial V}\right)_T \tag{24-27}$$

The first of these last three relations is used in the derivation of the equation of state from statistical mechanics.[25]

[25] See J. de Boer [*32*]; see also J. O. Hirschfelder, C. F. Curtiss, and R. B. Bird [*34*].

B,25. The Evaluation of the Thermodynamic Properties of Ideal Gases. In this article we discuss in detail an important application of the principles of statistical thermodynamics, namely the calculation of the thermodynamic properties of gases at sufficiently low densities that their behavior can be considered to be ideal. It is first shown that the symmetry restrictions imposed on the wave functions for systems made up of identical particles lead to two types of distributions of energy among the molecules in the gas: one for Fermi-Dirac gases and another for Bose-Einstein gases. At high temperatures, the two distribution functions approach one another, and the thermodynamic properties of the two kinds of gases become the same. This limiting form for the two types of statistics is referred to as "Boltzmann Statistics." In the latter portion of this article, the thermodynamic properties of Boltzmann gases are calculated and the contributions due to the internal degrees of freedom are discussed.

The partition function for an ideal gas. If the density of a gaseous system of N molecules is sufficiently low and if the intermolecular forces are short range (compared to Coulombic), the amount of time any molecule spends in collisions is negligible compared with the time between collisions. Under such conditions one may say that the intermolecular potential energy is negligible with respect to the total energy of the system and hence that the total energy is just the sum of the energies of the individual molecules. Each of the molecules may be in any one of the quantum states of the free molecule. However, since the molecules are identical, the state of the gas is specified by the number of molecules in each state, and it is meaningless to specify which molecules are in which states. That is, the kth state of the gas with energy E_k is specified by a set of occupation numbers, n_j^k, which give the number of molecules in the jth molecular quantum state.

Let the energy of a molecule in the jth state be ϵ_j. Then since the energy of the gas is the sum of that of the individual molecules, the energy of the gas in the kth state is

$$E_k = \sum_j n_j^k \epsilon_j \tag{25-1}$$

Clearly, the n_j^k satisfy the relation

$$N = \sum n_j^k \tag{25-2}$$

There is in addition a possible restriction on the n_j^k due to statistics. The occupation numbers are restricted to 0 and 1 for Fermi-Dirac statistics. However, for Bose-Einstein particles the symmetry requirements on the wave function of the gas do not limit the number of molecules in any state.

The partition function for the entire gas may be written as

$$Q_N = \sum_k e^{-E_k/kT}$$

$$= \sum_k e^{-(\Sigma_j n_j^k \epsilon_j)/kT}$$

$$= \sum_k w_1^{n_1^k} w_2^{n_2^k} w_3^{n_3^k} \cdots \qquad (25\text{-}3)$$

in which

$$w_j = e^{-\epsilon_j/kT} \qquad (25\text{-}4)$$

It is understood that the sum k is taken only over those sets of n_j^k which are consistent with the statistics of the individual molecules, and the constraint Eq. 25-2.

Let us now define a generating function $f(\zeta)$ by

$$f(\zeta) = \prod_j (1 \pm \zeta w_j)^{\pm 1} \qquad (25\text{-}5)$$

In this expression and those which follow, the upper sign corresponds to Fermi-Dirac statistics and lower sign to Bose-Einstein statistics. The partition function, Q_N, is then just the coefficient of ζ^N in the expansion of $f(\zeta)$. If ζ is now taken to be a complex variable, we can divide $f(\zeta)$ by ζ^{N+1} and use the method of residues (or Cauchy's theorem) to obtain Q_N:

$$Q_N = \frac{1}{2\pi i} \oint_C \zeta^{-N-1} f(\zeta) d\zeta \qquad (25\text{-}6)$$

The closed contour, C, selected for the integration in the complex domain must be chosen in such a way as to enclose $\zeta = 0$. This integral can be evaluated by the "method of steepest descents" and the result is

$$\ln Q_N = -N \ln \zeta_0 \pm \sum_j \ln (1 \pm \zeta_0 e^{-\epsilon_j/kT}) \qquad (25\text{-}7)$$

in the limit that N is large (i.e. a large sample with negligible surface effects). The auxiliary relation

$$N = \sum_j (\zeta^{-1}{}_0 e^{\epsilon_j/kT} \pm 1)^{-1} \qquad (25\text{-}8)$$

serves to define the parameter ζ_0 which is a positive real number.

Distribution of energy among the molecules of an ideal gas. Let us now consider the energy distribution among the molecules of a Fermi-Dirac and Bose-Einstein gas. As mentioned above, the probability that the system is in state k is proportional to $e^{-E_k/kt}$. Hence the probable number of molecules in state j is given by

$$\overline{n_j} = \frac{\sum_k n_j^k e^{-E_k/kT}}{\sum_k e^{-E_k/kT}} \tag{25-9}$$

When the quantities E_k are expressed by means of Eq. 25-1

$$\overline{n_j} = \frac{\sum_k n_j^k e^{-(\Sigma_i n_i^k \epsilon_i)/kT}}{Q_N} \tag{25-10}$$

so that one finally obtains

$$\overline{n_j} = -\frac{kT}{Q_N}\left(\frac{\partial Q_N}{\partial \epsilon_j}\right) \tag{25-11}$$

Care must be exercised in obtaining $(\partial Q_N/\partial \epsilon_j)$ from Eq. 25-7 for ζ_0 is a function of the quantities ϵ_j. The result of the differentiation is[26]

$$\overline{n_j} = (\zeta_0^{-1} e^{\epsilon_j/kT} \pm 1)^{-1} \quad \begin{array}{l} + \text{ sign: Fermi-Dirac} \\ - \text{ sign: Bose-Einstein} \end{array} \tag{25-12}$$

This describes the manner in which the energy in a Fermi-Dirac or Bose-Einstein gas is distributed among the individual molecules. The parameter ζ_0 may easily be shown to be related to the Gibbs free energy by $F = NkT \ln \zeta_0$.

For most actual systems $\zeta_0^{-1} \gg 1$, the main exceptions being electrons in metals and substances at extremely low temperatures or very high densities. To a good approximation $\zeta_0^{-1} = 3.122 \times 10^{-4} V_0(\mathfrak{M}T)^{\frac{3}{2}}$, where V_0 is the volume in cm^3 per mole and $\mathfrak{M}$ is the molecular weight. For a perfect gas at standard conditions $\zeta_0^{-1} = 31{,}590\mathfrak{M}^{\frac{3}{2}}$. Systems which have a value of ζ_0^{-1} small enough so that the ± 1 cannot be neglected are sometimes referred to as "degenerate" systems or are said to be "in a state of degeneracy." For most applications we can use the distribution

$$\overline{n_j} \cong \zeta_0 e^{-\epsilon_j/kT} \tag{25-13}$$

We see thus that the Fermi-Dirac and Bose-Einstein distributions approach as a limit the Maxwell-Boltzmann distribution. Furthermore, in this limit the partition function for the gas as given by Eq. 25-7 becomes

$$Q_N = \frac{Q^N}{N!} \tag{25-14}$$

where Q is the partition function for one molecule in the vessel

$$Q = \sum_j e^{-\epsilon_j/kT} \tag{25-15}$$

The $N!$ which appears here is of course due fundamentally to the indistinguishability of the molecules. Systems whose partition functions are given by Eq. 25-14 are said to obey Boltzmann statistics. The problem

[26] The distribution functions for Fermi-Dirac and Bose-Einstein gases may also be obtained by combinatorial analysis. See, for example, R. H. Tolman, [*33*, Chap. 10].

of calculating the thermodynamic properties for the Boltzmann gas has thus been reduced to that of determining the partition function for one molecule in the vessel. The computation of the partition function for a single molecule is considered in the remaining part of this article.

Contributions to the thermodynamic properties due to the translational and internal motions of the molecules. For an accurate calculation of the partition function for complex molecules it is necessary to know all of the energy levels of the system. Because of the large number of degrees of freedom this is very difficult, and a great deal of information is required which is ordinarily not available. For most practical purposes, however, it suffices to idealize the situation and to neglect the weak couplings between various degrees of freedom.

Let $\mathcal{H}$ be the quantum mechanical Hamiltonian of a complex molecule, and let the wave function and energy corresponding to the jth quantum state be ψ_j and ϵ_j respectively, so that

$$\mathcal{H}\psi_j = \epsilon_j\psi_j \tag{25-16}$$

For present purposes, the set of energy levels ϵ_j is taken relative to the ground state as zero. (In considering mixtures of different chemical species it is necessary to refer the energy of the ground states of chemical species to a consistent set of standard states chosen for the elements. This is discussed below under the subheading: *Ideal gas mixtures.*) The motions executed by polyatomic molecules are very complex. In general, however, it is possible to consider the motion as being made up of six types of motion: the over-all translation of the molecule (tr); the various stretching and bending vibrations (vib); the rotational motion of the entire molecule (rot); the various rotations or restricted rotations of groups within the molecule (int rot); the electronic motion (elec); and the nuclear spin (nucl). These several motions are to various extents independent of one another. For example, the rotational motion is nearly independent of the electronic motion but frequently rather strongly linked to the vibrational motion. Nevertheless as a first approximation we may assume that these motions are completely independent so that

$$\mathcal{H} = \mathcal{H}^{(\mathrm{tr})} + \mathcal{H}^{(\mathrm{rot})} + \sum \mathcal{H}^{(\mathrm{vib})} + \sum \mathcal{H}^{(\mathrm{int\ rot})} + \sum \mathcal{H}^{(\mathrm{elec})} + \sum \mathcal{H}^{(\mathrm{nucl})} \tag{25-17}$$

$$= \mathcal{H}^{(\mathrm{tr})} + \sum \mathcal{H}^{(\mathrm{i})} \tag{25-18}$$

Here we designate the various internal (nontranslational) motions by superscript (i). This assumption then enables one to separate the Schrödinger equation (25-16) into component parts:

$$\mathcal{H}^{(\mathrm{tr})}\phi_j^{(\mathrm{tr})} = \epsilon_j^{(\mathrm{tr})}\phi_j^{(\mathrm{tr})} \tag{25-19}$$

$$\mathcal{H}^{(\mathrm{i})}\phi_j^{(\mathrm{i})} = \epsilon_j^{(\mathrm{i})}\phi_j^{(\mathrm{i})} \tag{25-20}$$

The solution of these individual Schrödinger equations gives the energy levels necessary to obtain partition functions for the various types of motions

$$Q^{(\mathrm{tr})} = \sum_j e^{-\epsilon_j^{(\mathrm{tr})}/kT} \tag{25-21}$$

$$Q^{(\mathrm{i})} = \sum_j e^{-\epsilon_j^{(\mathrm{i})}/kT} \tag{25-22}$$

Since the energy of the molecule in any state is assumed to be simply the sum of the energies associated with the several types of motion, we write the molecular partition function as

$$Q = Q^{(\mathrm{tr})} \prod_{\mathrm{i}} Q^{(\mathrm{i})} \tag{25-23}$$

Now according to Eq. 25-14, the partition function for the entire gas of N molecules may be written as

$$Q_N = \frac{Q^N}{N!} = \frac{[Q^{(\mathrm{tr})}]^N}{N!} \left[\prod_{\mathrm{i}} Q^{(\mathrm{i})}\right]^N \tag{25-24}$$

As it is always $\ln Q_N$ which appears in the expressions for the thermodynamic functions, these may be written as a sum of the contributions from the translational and various internal motions

$$X = X^{(\mathrm{tr})} + \sum_{\mathrm{i}} X^{(\mathrm{i})} \tag{25-25}$$

in which X can be internal energy, specific heats, free energy, etc. It is customary to include the $N!$ in the translational contribution as indicated; the latter is the predominant contribution in most cases. Accordingly, the various contributions are given by[27]

$$\begin{aligned}
U^{(\mathrm{tr})} &= NkT^2\left(\frac{\partial \ln Q^{(\mathrm{tr})}}{\partial T}\right) & U^{(\mathrm{i})} &= NkT^2\left(\frac{\partial \ln Q^{(\mathrm{i})}}{\partial T}\right)\\
S^{(\mathrm{tr})} &= \frac{U^{(\mathrm{tr})}}{T} + Nk\left[\ln\left(\frac{Q^{(\mathrm{tr})}}{N}\right) + 1\right] & S^{(\mathrm{i})} &= \frac{U^{(\mathrm{i})}}{T} + Nk \ln Q^{(\mathrm{i})}\\
A^{(\mathrm{tr})} &= -NkT\ln\left(\frac{Q^{(\mathrm{tr})}}{N}\right) - NkT & A^{(\mathrm{i})} &= -NkT \ln Q^{(\mathrm{i})}\\
H^{(\mathrm{tr})} &= U^{(\mathrm{tr})} + NkT & H^{(\mathrm{i})} &= U^{(\mathrm{i})}\\
F^{(\mathrm{tr})} &= A^{(\mathrm{tr})} + NkT & F^{(\mathrm{i})} &= A^{(\mathrm{i})}\\
C_v^{(\mathrm{tr})} &= (\partial U^{(\mathrm{tr})}/\partial T)_v & C_v^{(\mathrm{i})} &= (\partial U^{(\mathrm{i})}/\partial T)_v\\
C_p^{(\mathrm{tr})} &= C_v^{(\mathrm{tr})} + Nk & C_p^{(\mathrm{i})} &= C_v^{(\mathrm{i})}
\end{aligned} \tag{25-26}$$

[27] In the translational contributions, Stirling's formula for $\ln N!$ has been used:

$$\ln N! \cong N \ln N - N$$

Thus to calculate the thermodynamic functions of an ideal gas, one simply adds together the contributions for the various types of motion which are active within the molecules. A complete description of the methods for calculating these various contributions for complex molecules would require a discussion of great length and the inclusion of many tabulated functions. Inasmuch as such complete treatments of the subject are readily available in several standard references,[28] the discussion here is confined to several simple calculations for monatomic and diatomic molecules.

Translational contributions. Unless the gas is at a temperature very close to absolute zero, the translational contributions to the thermodynamic properties may be computed from the classical partition function. The molecule is considered to be confined to a volume V and its classical Hamiltonian is simply $(p_x^2 + p_y^2 + p_z^2)/2m$. Then the phase integral for the translational motion is

$$Q^{(tr)} = \frac{V}{h^3} \int_{-\infty}^{+\infty} \int_{-\infty}^{+\infty} \int_{-\infty}^{+\infty} e^{-\frac{p_x^2+p_y^2+p_z^2}{2mkT}} dp_x dp_y dp_z \qquad (25\text{-}27)$$

in which the factor V results from the triple integration over the spatial coordinates x, y, z. The integration over the momenta is a standard integral, giving

$$Q^{(tr)} = \frac{V}{h^3} (2\pi mkT)^{\frac{3}{2}} \qquad (25\text{-}28)$$

This shows explicitly how the partition function can depend on certain mechanical parameters of the system—in this case, the volume. From formulas relating the thermodynamic functions to the partition function the familiar relations for the quantities per mole,

$$U^{(tr)} = \tfrac{3}{2}RT; \qquad C_V^{(tr)} = \tfrac{3}{2}R; \qquad C_p^{(tr)} = \tfrac{5}{2}R \qquad (25\text{-}29)$$

can be easily verified. It may also be shown that the entropy per mole is

$$\begin{aligned} S^{(tr)} &= R(\tfrac{3}{2} \ln M + \tfrac{3}{2} \ln T + \ln V) + 2.6546 \\ &= R(\tfrac{3}{2} \ln M + \tfrac{5}{2} \ln T - \ln p) - 2.3141 \end{aligned} \qquad (25\text{-}30)$$

in which the constants are just combinations of various universal constants.[29] This expression for the entropy is the Sackur-Tetrode equation. For a monatomic gas, the translational contributions are the sole con-

[28] The most usable treatment is that given by J. E. Mayer and M. G. Mayer [*36*].

[29] These values for the constants are correct if R is in cal/mole degree, T is degrees K, V is liters, and p is atmospheres. Throughout the section we use the defined calorie, 4.1833 international Joules or 4.18401 absolute Joules as recommended by the National Bureau of Standards. Here we take $R = 1.98718$ cal/mole deg.

tributions to the thermodynamic properties. For gases composed of polyatomic molecules these are generally the primary contributions.

Rotational contributions in diatomic molecules. If a diatomic molecule is pictured as a rigid dumbbell with moment of inertia I rotating in space, solution of the Schrödinger equation for the system gives the energy and degeneracy of the Jth level, where J is the rotational quantum number

$$\epsilon_J^{(\mathrm{rot})} = J(J+1)(\hbar^2/2I) \tag{25-31}$$

$$g_J^{(\mathrm{rot})} = 2J+1 \tag{25-32}$$

Thus the partition function is

$$Q^{(\mathrm{rot})} = \sum_{J=0}^{\infty} (2J+1)e^{-\frac{\hbar^2 J(J+1)}{2IkT}} \tag{25-33}$$

The moment of inertia can be obtained from spectroscopic data. The Euler-Maclaurin summation formula is very useful for the evaluation of such expressions

$$\sum_{J=J_0}^{J=J_1} f(J) = \int_{J_0}^{J_1} f(J)dJ + \frac{1}{2}[f(J_0) + f(J_1)]$$
$$+ \sum_{k=1}^{\infty} (-1)^k \frac{B_k}{(2k)!}\left[\left(\frac{d^{2k-1}f}{dJ^{2k+1}}\right)_{J_0} - \left(\frac{d^{2k-1}f}{dJ^{2k-1}}\right)_{J_1}\right] \tag{25-34}$$

Here the B_k are the Bernoulli numbers: $B_1 = \frac{1}{6}$, $B_2 = \frac{1}{30}$, $B_3 = \frac{1}{42}$, $B_4 = \frac{1}{30}$, $B_5 = \frac{5}{66}$, · · · . Substituting

$$(2J+1)e^{-\hbar^2 J(J+1)/2IkT}$$

for $f(J)$ and letting $J_0 = 0$ and $J_1 = \infty$ we obtain

$$Q^{(\mathrm{rot})} = \frac{2IkT}{\hbar^2} + \frac{1}{3} + \frac{1}{15}\left(\frac{\hbar^2}{2IkT}\right) + \cdots \tag{25-35}$$

But for diatomic molecules at normal temperatures $(2IkT/\hbar^2) \gg 1$, so that only the first term is important.

It is usual to include in the denominator of the rotational partition function a symmetry number σ, which is 1 for heteronuclear and 2 for homonuclear diatomic molecules. This factor arises from the restrictions on the number of allowed quantum states imposed by the Pauli exclusion principle.

The first term of Eq. 25-35 (aside from the symmetry number) may be obtained from the evaluation of the classical phase integral. From these results one can obtain the following expressions for the thermodynamic properties per mole:

$$U^{(\text{rot})} = H^{(\text{rot})} = RT \tag{25-36}$$

$$C_p^{(\text{rot})} = C_V^{(\text{rot})} = R \tag{25-37}$$

$$S^{(\text{rot})} = R\left[1 + \ln \frac{2IkT}{\sigma\hbar^2}\right] \tag{25-38}$$

Vibrational contributions in diatomic molecules. If in a diatomic molecule the stretching of the bond is assumed to obey Hooke's law (the force tending to restore the molecule to its equilibrium position is directly proportional to the distortion of the bond) then the problem reduces itself to the solution of the Schrödinger equation for the one-dimensional simple harmonic oscillator. The energy level of the nth vibrational state is

$$\epsilon_n^{(\text{vib})} = h\nu(n + \tfrac{1}{2}) \tag{25-39}$$

These vibrational states are all nondegenerate. The vibrational partition function is usually defined in terms of energy above the zero-point energy, so that

$$Q^{(\text{vib})} = \sum_{n=0}^{\infty} e^{-nh\nu/kT} = \frac{1}{1 - e^{-h\nu/kT}} \tag{25-40}$$

The last step is achieved by using the formula for the sum of a geometric series. The associated thermodynamic contributions per mole are then

$$U^{(\text{vib})} = H^{(\text{vib})} = Nh\nu(e^{h\nu/kT} - 1)^{-1} \tag{25-41}$$

$$S^{(\text{vib})} = \frac{U^{(\text{vib})}}{T} - R \ln (1 - e^{-h\nu/kT}) \tag{25-42}$$

The frequency ν is the natural vibration frequency of the molecule and is obtained from vibration spectra. At high temperatures, corrections for the increasing anharmonicity of the oscillations must be applied.

Electronic contributions. In most molecules, rather high temperatures are required for the activation of the higher electronic levels. Consequently, in most cases $Q^{(\text{elec})}$ is simply the degeneracy of the electronic ground state. The electronic degeneracy for monatomic, diatomic, and polyatomic gases may be obtained from the spectroscopist's "term values" in the following manner:

1. Monatomic Molecules. In the term symbols for the electronic state of the atom (e.g. $^3P_3$, $^1S_2$, $^4F_0$), the right subscript indicates the total (orbital + spin) angular momentum J of all the electrons in the atom. The electronic degeneracy is given by:

$$g^{(\text{elec})} = 2J + 1 \tag{25-43}$$

2. Diatomic Molecules. In the term symbols for the electronic states of the diatomic molecules (e.g. $^3\Sigma_g^-$, $^2\Pi_u$, $^1\Delta_g$), the left superscript indicates the "spin multiplicity," $2S + 1$. The electronic degeneracy is then:

$$(2S + 1) \quad \text{for } \Sigma \text{ states}$$
$$2(2S + 1) \quad \text{for } \Pi, \Delta, \text{etc. states}$$

3. Polyatomic Molecules. In polyatomic valence-saturated molecules, $g^{(\text{elec})} = 1$ or 2 according to whether there are an even or odd number of electrons (since for an even number of electrons there is no resultant spin angular momentum, but for an odd number of electrons there is a resultant spin of $\frac{1}{2}$).

Nuclear spin contributions. The nuclear spin partition function is simply the product of the nuclear spin multiplicities for all the atoms in the molecule

$$Q^{(\text{nucl})} = g^{(\text{nucl})} = \prod_j (2S_j^{(\text{nucl})} + 1) \tag{25-44}$$

where $S_j^{(\text{nucl})}$ is the nuclear spin of the jth atom. However, since this contribution affects only the additive constant on the entropy and always cancels out (except in processes involving transmutation of the elements) it is usually ignored. The result is "virtual" entropies rather than "absolute" entropies.

Ideal gas mixtures. It can be shown that for an ideal gas mixture containing N_A molecules of substance A and N_B molecules of B, $(N = N_A + N_B)$, the partition function is

$$Q_N = \frac{Q_A^{N_A} Q_B^{N_B}}{N_A! N_B!} e^{-\frac{N_A(\epsilon_0)_A + N_B(\epsilon_0)_B}{kT}} \tag{25-45}$$

Here the Q_A and Q_B are the molecular partition functions discussed above in which all the energies are referred to the ground states of the respective molecules. The factors $N_A!$ and $N_B!$ arise because of the indistinguishability of molecules of the same species in the same way that the $N!$ arises in the corresponding expression for the partition function of a pure substance. The $(\epsilon_0)_A$ and $(\epsilon_0)_B$ are the energies of the ground states of molecules A and B referred to a consistent set of standard states for the elements. Thus $(\epsilon_0)_A$ is the energy of formation of a molecule of A in the ground state from the elements of which it is composed, each being in its own standard state. The standard states of the elements are usually taken to be the elements in the form in which they naturally occur at zero degrees centigrade.

From the above expression (Eq. 25-45) for the partition function it is possible to calculate the various thermodynamic properties for mixtures of ideal gases. It is thus found that

$$U_{\text{mixture}} = \sum X_j U_j \tag{25-46}$$

$$H_{\text{mixture}} = \sum X_j H_j \tag{25-47}$$

$$C_{p_{\text{mixture}}} = \sum X_j C_{p_j} \tag{25-48}$$

This last expression is valid only for nonreacting mixtures. The entropy for a mixture of gases is not simply the sum of contributions from the individual components. Substitution of the partition function for a mixture into Eq. 24-18 for the entropy shows that the entropy of a mixture is given by

$$S_{\text{mixture}} = \sum X_j S_j - Nk \sum X_j \ln X_j \tag{25-49}$$

The quantity $-Nk\sum X_j \ln X_j$ is referred to as the entropy of mixing. Similar contributions due to mixing arise in the expressions for the free energy F and the work function A.

In treating the temperature dependence of the thermodynamical properties of a mixture, consideration must be given as to whether chemical reactions can be ignored or whether chemical equilibrium is maintained between specified molecular species. In the first case, the numbers N_A and N_B remain constant, whereas in the second case they vary in accordance with the equilibrium constants.

CHAPTER 5. THE KINETIC THEORY OF GASES[30]

J. O. HIRSCHFELDER C. F. CURTISS
R. B. BIRD E. L. SPOTZ

The preceding chapter is concerned with the properties of systems in equilibrium. In this chapter the statistical methods are applied to the description of the nonequilibrium properties—in particular the transport phenomena. In this chapter we arrive at expressions for the hydrodynamical equations and formulas for the transport coefficients. Sec. D deals with the evaluation and the practical computation of the transport coefficients.

We begin by showing that the properties of a dilute gas are completely described by the distribution function $f(\mathbf{r}, \mathbf{v}, t)$, defined so that $f(\mathbf{r}, \mathbf{v}, t)d\mathbf{r}d\mathbf{v}$ is the probable number of molecules which at time t have position coordinates $\mathbf{r}$ between $\mathbf{r}$ and $\mathbf{r} + d\mathbf{r}$ and have a velocity $\mathbf{v}$ be-

[30] This chapter is based upon material given in Chapter 7 of *Molecular Theory of Gases and Liquids*, by J. O. Hirschfelder, C. F. Curtiss, and R. B. Bird (John Wiley & Sons, Inc., 1954). The authors wish to thank the publisher for permission to use this material.

tween $\mathbf{v}$ and $\mathbf{v} + d\mathbf{v}$. This distribution function is given as the solution of an integro-differential equation, known as the Boltzmann equation. This equation is valid at densities sufficiently low that the effect of collisions involving more than two molecules is negligible. If the mean free path of the molecules in the gas is short compared with all of the macroscopic dimensions, the gas behaves as a continuum. Under such conditions the Boltzmann equation leads to the Navier-Stokes equations of hydrodynamics and expressions for the flux vectors. The transport coefficients are defined in terms of the flux vectors. Expressions for the transport coefficients are obtained in terms of integrals involving the intermolecular potential function.

B,26. The Distribution Functions. The kinetic theory in this chapter is based on the Boltzmann equation which specifies the distribution function $f(\mathbf{r}, \mathbf{v}, t)$. We give here two derivations of the Boltzmann equation. The first, a simple physical derivation, gives physical interpretation to the various terms. The second, a more rigorous derivation, is based on the integration of the Liouville equation. Before presenting these derivations, we discuss some of the properties of the various kinetic theory distribution functions and their use in describing nonequilibrium systems.

Physical description of nonequilibrium systems.[31] The exact dynamical state of a system of particles is given by specifying the complete set of position and momentum coordinates of all the individual particles. According to the laws of classical mechanics a knowledge of the exact dynamical state at a particular time permits an exact prediction of the state at any future time.

It is virtually impossible to give a complete description of the state of a complex macroscopic system. We must content ourselves with descriptions of the system which are considerably less than complete. The problem of predicting the probable behavior of a system from incomplete information about the system at some specified time is a statistical one. It is useful to employ the technique of representing the system by means of an ensemble consisting of a large number of replicas of the single system. The state of the ensemble is then described by a distribution function $f^{(N)}(\mathbf{r}^N, \mathbf{p}^N, t)$ in the phase space of a single system. This distribution function is so chosen that averages over the ensemble are in exact agreement with the incomplete (macroscopic) knowledge of the state of the system at some specified time. Then the probable behavior of the system at subsequent times is taken to be the average behavior of members of the representative ensemble. There are, of course, many different ways in which the ensemble could be formed, and therefore the distribution function is not uniquely specified. In the equilibrium case, the ensemble is specified by the ergodic theorem of Birkhoff. Such

[31] The discussion presented here is similar to that of H. Grad [*37*].

a theorem has not yet been developed in the nonequilibrium case. However, this offers no difficulties, since the study of nonequilibrium statistical mechanics is principally concerned with the lower order distribution functions, $f^{(1)}$ and $f^{(2)}$. Equations for the lower order distribution functions are derived by introducing restrictions, such as the concept of "molecular chaos," which effectively restrict the consideration to certain types of distribution functions $f^{(N)}$.

The variation of the distribution function $f^{(N)}(\mathbf{r}^N, \mathbf{p}^N, t)$ with time is described by the Liouville equation. This equation, involving $6N$ variables, is difficult to solve. Fortunately, one is usually not interested in a description of the system as complete as that afforded by $f^{(N)}$. Rather, one is satisfied with the less complete description given by one of the lower order distribution functions, $f^{(h)}(\mathbf{r}^h, \mathbf{p}^h, t)$. These functions are obtained by integrating $f^{(N)}$ over the coordinates of the $N - h$ molecules not included in the group $1, 2, 3, \ldots, h$. Of particular interest are the distribution functions with $h = 1$ and $h = 2$.

In this chapter we are concerned primarily with the function $f^{(1)}(\mathbf{r}, \mathbf{p}, t)$ which gives the probability of finding one particular molecule with the specified position and momentum. The configuration and momenta of the other $N - 1$ molecules remain unspecified. In considering a system made up of identical molecules, the distribution function $f^{(N)}$ is symmetric in the coordinates of all the molecules, inasmuch as no physical experiment differentiates among them. Consequently, in obtaining $f^{(1)}$ it is immaterial which molecule is singled out as special. Clearly, $f^{(1)}$ is adequate for the description of all of the physical properties of gases which do not depend upon the relative positions of two or more molecules. This means that the level of information corresponding to $f^{(1)}$ is sufficient for studying the behavior of moderately dilute gases.

For gases at higher density, a knowledge of higher order distribution functions is required. If, however, two-body forces can be assumed (i.e. $\Phi(\mathbf{r}^N) = \frac{1}{2}\sum_{i,j}\phi_{ij}$), the distribution function of order $h = 2$ is sufficient to determine all of the macroscopic properties of the system. The distribution function $f^{(2)}(\mathbf{r}_1, \mathbf{r}_2, \mathbf{p}_1, \mathbf{p}_2, t)$ is the distribution in the phase space of pairs. This pair distribution function is used in the study of the behavior of dense gases.

Let us now consider the time-dependence of the various distribution functions. As already mentioned, the manner in which $f^{(N)}$ changes with time is given by the Liouville equation. That is, to each $f^{(N)}$ at an initial time t_0 there corresponds uniquely a function $f^{(N)}$ at a later time t_1. However, for the lower order distribution functions, it is not possible to predict from a knowledge of $f^{(h)}(t_0)$ a unique value of $f^{(h)}(t_1)$. For example, at t_0 it is possible that there are a number of functions $f^{(N)}(t_0)$, $f^{(N)\prime}(t_0)$, $f^{(N)\prime\prime}(t_0)$, . . . which, when integrated over the variables corresponding to $N - 1$ molecules, all give the same function $f^{(1)}(t_0)$. Later on the group of Nth order functions becomes $f^{(N)}(t_1)$, $f^{(N)'}(t_1)$, $f^{(N)''}(t_1)$, . . . , and there correspond to these functions the singlet distribution functions $f^{(1)}(t_1)$, $f^{(1)'}(t_1)$, $f^{(1)''}(t_1)$, . . . which are in general different from one another. This means that no unique integro-differential equation exists for $f^{(1)}$. In order to remove this ambiguity it is necessary to invoke an additional condition which restricts the possible functions $f^{(N)}$. This is the condition of *molecular chaos*, which is introduced into the derivation of the Boltzmann equation for $f^{(1)}$.

A physical derivation of the Boltzmann equation. Let us consider a monatomic gas mixture in a nonequilibrium state. The gas is assumed to be sufficiently dilute that two-body but not three-body collisions are important. For generality we suppose that the molecules of the ith species are subject to an external force, $\mathbf{X}_i$. In this treatment $\mathbf{X}_i$ may be a function of position and time, but not of velocity. The effect of velocity-dependent forces is considered in a somewhat different manner [*38*]. The external force is assumed to be much smaller than the forces which act on the molecules during an encounter. The intermolecular forces are generally of the order of many powers of ten times the force of gravity and act only during the very short time of encounter.

As discussed above we are interested in the description of the gas in terms of the distribution function in the phase space of a single molecule (μ space). In the case of a mixture there is a distribution function $f_i^{(1)}(\mathbf{r}, \mathbf{p}_i, t)$ for each component such that the probable number of molecules of kind i with position coordinates in the range $d\mathbf{r}$ about $\mathbf{r}$ and with momentum coordinates in the range $d\mathbf{p}_i$ about $\mathbf{p}_i$ is $f_i^{(1)}(\mathbf{r}, \mathbf{p}_i, t)d\mathbf{r}d\mathbf{p}_i$. According to equilibrium statistical mechanics the function $f_i^{(1)}$ at equilibrium is independent of time and space and the velocity distribution is Maxwellian. Now it is desired to ascertain the manner in which $f_i^{(1)}(\mathbf{r}, \mathbf{p}_i, t)$ depends upon the variables in nonequilibrium situations. That is, we wish to determine the nature of the flow of phase points through a six-dimensional phase space, where each phase point represents one molecule and the molecules interact with one another. If there were no interaction between the individual molecules, the behavior of the function $f_i^{(1)}(\mathbf{r}, \mathbf{p}_i, t)$ would be given by a Liouville equation for points in μ space. We shall find, in fact, that the assumption of interaction simply modifies the Liouville equation by the addition of terms which account for molecular collisions.

We visualize a region of volume $d\mathbf{r}d\mathbf{p}_i$ located about a point $\mathbf{r}$, $\mathbf{p}_i$. In this element of volume there are $f_i^{(1)}d\mathbf{r}d\mathbf{p}_i$ phase points associated with particles of the ith kind. In the absence of collisions in the gas, the molecules corresponding to these phase points move in such a way that at time $t + dt$, their position vectors[32] are $[\mathbf{r} + (\mathbf{p}_i/m_i)dt]$ and their momentum vectors are $[\mathbf{p}_i + \mathbf{X}_i dt]$. No other phase points arrive at this

[32] Here we use explicitly the properties of Cartesian coordinate systems.

latter position (in absence of collisions), so that

$$f_i^{(1)}(\mathbf{r}, \mathbf{p}_i, t)d\mathbf{r}d\mathbf{p}_i = f_i^{(1)}\left[\left(\mathbf{r} + \frac{\mathbf{p}_i}{m_i}\,dt\right), (\mathbf{p}_i + \mathbf{X}_i dt), (t + dt)\right] d\mathbf{r}d\mathbf{p}_i \quad (26\text{-}1)$$

But since collisions are taking place in the gas, not all of the phase points at $[\mathbf{r}, \mathbf{p}_i]$ arrive at

$$\left[\left(\mathbf{r} + \frac{\mathbf{p}_i}{m_i}\,dt\right), (\mathbf{p}_i + \mathbf{X}_i dt)\right]$$

after the interval dt, for the molecules associated with these phase points which are deflected from their course by collisions suffer changes in momentum. There are also some phase points which did not begin at $\mathbf{p}$, $\mathbf{r}_i$ but which, as a result of colliding with other molecules, arrive at

$$\left[\left(\mathbf{r} + \frac{\mathbf{p}_i}{m_i}\,dt\right), (\mathbf{p}_i + \mathbf{X}_i dt)\right]$$

Let the number of molecules of the ith kind, lost from the momentum range $\mathbf{p}_i$ to $\mathbf{p}_i + d\mathbf{p}_i$ in the position range $\mathbf{r}$ and $\mathbf{r} + d\mathbf{r}$ because of collisions with molecules of type j during the time interval dt, be $\Gamma_{ij}^{(-)}d\mathbf{r}d\mathbf{p}_i dt$. Similarly, the number of molecules of the ith kind which in a time dt join the group of points which started from $[\mathbf{r}, \mathbf{p}_i]$ because of collision with molecules of type j is denoted by $\Gamma_{ij}^{(+)}d\mathbf{r}d\mathbf{p}_i dt$. When the equation for the flow of phase points takes into account the effects of collisions, it becomes

$$f_i^{(1)}\left[\left(\mathbf{r} + \frac{\mathbf{p}_i}{m_i}\,dt\right), (\mathbf{p}_i + \mathbf{X}_i dt), (t + dt)\right] d\mathbf{r}d\mathbf{p}_i$$
$$= f_i^{(1)}(\mathbf{r}, \mathbf{p}_i, t)d\mathbf{r}d\mathbf{p}_i + \sum_j [\Gamma_{ij}^{(+)} - \Gamma_{ij}^{(-)}]d\mathbf{r}d\mathbf{p}_i dt \quad (26\text{-}2)$$

The term on the left-hand side of this equation may be expanded in a Taylor series about the point $\mathbf{r}$, $\mathbf{p}_i$, t,

$$f_i^{(1)}\left[\left(\mathbf{r} + \frac{\mathbf{p}_i}{m_i}\,dt\right), (\mathbf{p}_i + \mathbf{X}_i dt), (t + dt)\right] d\mathbf{r}d\mathbf{p}_i$$
$$= \left[f_i^{(1)}(\mathbf{r}, \mathbf{p}_i, t) + \frac{1}{m_i}(\mathbf{p}_i \cdot \nabla f_i^{(1)})dt + (\mathbf{X}_i \cdot \nabla_{p_i} f_i^{(1)})dt + \left(\frac{\partial f_i^{(1)}}{\partial t}\right)dt + \cdots\right] d\mathbf{r}d\mathbf{p}_i \quad (26\text{-}3)$$

These two equations may be combined to give the Boltzmann equation:

$$\frac{\partial f_i^{(1)}}{\partial t} + \frac{1}{m_i}(\mathbf{p}_i \cdot \nabla f_i^{(1)}) + (\mathbf{X}_i \cdot \nabla_{p_i} f_i^{(1)}) = \sum_j (\Gamma_{ij}^{(+)} - \Gamma_{ij}^{(-)}) \quad (26\text{-}4)$$

which describes the time rate of change of the function $f_i^{(1)}$. It has the same general form as the Liouville equation except for the addition of the collision terms on the right-hand side.

It should be mentioned that the quantities $\Gamma_{ij}^{(+)}$ and $\Gamma_{ij}^{(-)}$ do not include the contributions resulting from the molecules colliding with the walls. These contributions are taken into account in the boundary conditions which are imposed on the hydrodynamical equations.

An explicit expression for the terms on the right-hand side of the Boltzmann equation—the "collision integrals"—can be found from the following arguments: First we examine $\Gamma_{ij}^{(-)}$, the number of molecules of type i which are removed from the volume element $d\mathbf{r}d\mathbf{p}_i$ by collisions with the molecules of type j during an element of time dt. Let us consider

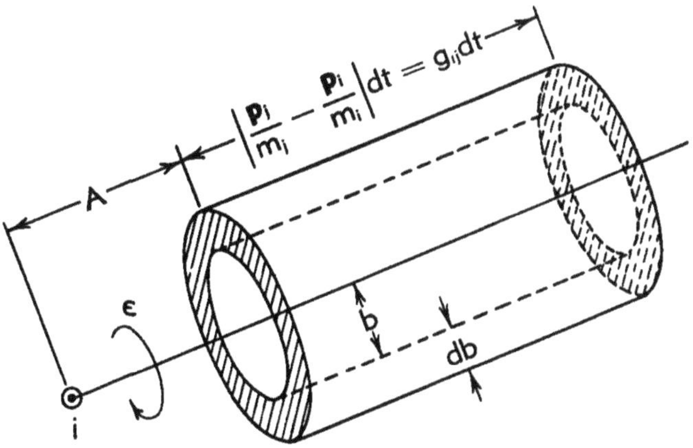

Fig. B,26. Collisions of molecules of type "j" with one molecule of type "i."

a molecule of type i, located at the position $\mathbf{r}$, and having a momentum $\mathbf{p}_i$. We wish to find the probability that this molecule will experience a collision with a molecule of type j, in the time interval dt, with the impact parameter[33] in a range db about b. If molecule "i" is considered to be fixed, then molecule "j" approaches it with a relative velocity $[(\mathbf{p}_j/m_j) - (\mathbf{p}_i/m_i)]$. This way of looking at the collision is pictured in Fig. B,26. It is assumed that the intermolecular force is negligible for distances of separation greater than a distance A, which is small compared to the mean free path.

From Fig. B,26 it is seen that any molecule of type j within the cylindrical shell will undergo an encounter with molecule i during the time interval dt—an encounter characterized by an impact parameter b

[33] The impact parameter b is the distance of closest approach of the two molecules if they continued to move in straight lines with their initial velocities and were not acted upon by intermolecular forces.

and an initial relative velocity

$$\mathbf{g}_{ji} = \frac{\mathbf{p}_j}{m_j} - \frac{\mathbf{p}_i}{m_i} \tag{26-5}$$

This picture is useful for visualizing collisions with molecule i by molecules of type j under conditions where the impact parameter is b, and the relative velocity is $\mathbf{g}_{ij}$. The distance A is essentially the intermolecular distance at which the potential begins to "take hold." During a short time interval dt any molecule of type j, which is located in the cylinder of base $bdbd\epsilon$ and height $g_{ij}dt$, will begin to undergo a collision with the molecule i. The probable number of molecules of type j within this cylinder is

$$f_j^{(1)}(\mathbf{r}, \mathbf{p}_j, t)g_{ij}bdbd\epsilon dt \tag{26-6}$$

where

$$g_{ij} = g_{ji} = |\mathbf{g}_{ji}| \tag{26-7}$$

Now the total number of collisions experienced by this molecule i with molecules of type j is obtained by adding together the number of collisions characterized by all values of the parameters b and ϵ and all relative velocities $\mathbf{g}_{ji}$. The result is

$$dt \iiint f_j^{(1)}(\mathbf{r}, \mathbf{p}_j, t)g_{ij}bdbd\epsilon d\mathbf{p}_j \tag{26-8}$$

Since the probable number of molecules of type i in the volume element $d\mathbf{r}$ about $\mathbf{r}$ with momentum in the range $d\mathbf{p}_i$ about $\mathbf{p}_i$ is $f_i^{(1)}(\mathbf{r}, \mathbf{p}_i, t)d\mathbf{r}d\mathbf{p}_i$, it follows that

$$\Gamma_{ij}^{(-)}d\mathbf{r}d\mathbf{p}_i dt = d\mathbf{r}d\mathbf{p}_i dt \iiint f_j^{(1)}(\mathbf{r}, \mathbf{p}_j, t)f_i^{(1)}(\mathbf{r}, \mathbf{p}_i, t)g_{ij}bdbd\epsilon d\mathbf{p}_j \tag{26-9}$$

Hence

$$\Gamma_{ij}^{(-)} = \iiint f_i^{(1)}f_j^{(1)}g_{ij}bdbd\epsilon d\mathbf{p}_j \tag{26-10}$$

represents the contribution to $\partial f_i^{(1)}/\partial t$ due to the removal from the group of molecules i by collisions with molecules of j. (In the last equation $f_i^{(1)}$ and $f_j^{(1)}$ are abbreviations for $f_i^{(1)}(\mathbf{r}, \mathbf{p}_i, t)$ and $f_j^{(1)}(\mathbf{r}, \mathbf{p}_j, t)$.)

The remaining portion of the collision integral $\Gamma_{ij}^{(+)}$ may be evaluated in a similar manner. Let us consider a collision between molecules with momenta $\mathbf{p}_i$ and $\mathbf{p}_j$ characterized by an impact parameter b. The momenta of the molecules after collision are denoted by $\mathbf{p}_i'$ and $\mathbf{p}_j'$. The values of the momenta after the collision are determined by the values of the momenta before the collision, the collision parameters, and the nature of the intermolecular force. For potential functions which are spherically symmetric it follows from the conservation of energy that the absolute values of the relative velocity before and after the collision are equal,

$$g_{ij} = g_{ij}' \tag{26-11}$$

And from the conservation of angular momentum the impact parameters before and after the collisions are equal,

$$b = b' \tag{26-12}$$

Because of the symmetry of the dynamical equations, a collision, with impact parameter b, between molecules with momenta $\mathbf{p}'_i$ and $\mathbf{p}'_j$ leaves the molecules with momenta $\mathbf{p}_i$ and $\mathbf{p}_j$.

By arguments identical with those above it follows that the probable number of collisions in the range $d\mathbf{r}$ about $\mathbf{r}$ in the time interval dt which result in molecules with momenta in the range $d\mathbf{p}_i$ about $\mathbf{p}_i$ is

$$\Gamma_{ij}^{(+)} d\mathbf{r} d\mathbf{p}_i dt = d\mathbf{r} d\mathbf{p}'_i dt \iiint f_i^{(1)}(\mathbf{r}, \mathbf{p}'_i, t) f_j^{(1)}(\mathbf{r}, \mathbf{p}'_j, t) g'_{ij} b' db' d\epsilon d\mathbf{p}'_j \tag{26-13}$$

Here the primed quantities are functions of the unprimed quantities, the functional relationship being determined by the nature of the intermolecular force. It is a direct consequence of the Liouville theorem that

$$d\mathbf{p}'_i d\mathbf{p}'_j = d\mathbf{p}_i d\mathbf{p}_j \tag{26-14}$$

This fact may be used, along with the equivalence before and after collision of both g_{ij} and b, to obtain $\Gamma_{ij}^{(+)}$

$$\Gamma_{ij}^{(+)} = \iiint f_i^{(1)\prime} f_j^{(1)\prime} g_{ij} b db d\epsilon d\mathbf{p}_j \tag{26-15}$$

where $f_i^{(1)\prime}$ and $f_j^{(1)\prime}$ represent $f_i^{(1)}(\mathbf{r}, \mathbf{p}'_i, t)$ and $f_j^{(1)}(\mathbf{r}, \mathbf{p}'_j, t)$. This expression gives the contribution to $\partial f_i^{(1)}/\partial t$ due to additions to the group of molecules by collision processes.

Now the expressions for $\Gamma_{ij}^{(+)}$ and $\Gamma_{ij}^{(-)}$ are substituted into the Boltzmann equation as it appears in Eq. 26-4, and the following equation for $f_i^{(1)}(\mathbf{r}, \mathbf{p}_i, t)$ results:

$$\frac{\partial f_i^{(1)}}{\partial t} + \frac{1}{m_i} (\mathbf{p}_i \cdot \nabla f_i^{(1)}) + (\mathbf{X}_i \cdot \nabla_{p_i} f_i^{(1)}) = \sum_j \iiint (f_i^{(1)\prime} f_j^{(1)\prime} - f_i^{(1)} f_j^{(1)}) g_{ij} b db d\epsilon d\mathbf{p}_j \tag{26-16}$$

This is the important Boltzmann integro-differential equation for the distribution function. Such an equation may be written for all the components in the gas mixture. In each of these equations, the distribution functions for all of the components appear on the right-hand side of the equation under the integral sign. It should be kept in mind that the law of force enters these integrals implicitly. The functions $f_i^{(1)\prime}$ and $f_j^{(1)\prime}$ are functions of $\mathbf{p}'_i$ and $\mathbf{p}'_j$ which can be calculated from dynamical principles when $\mathbf{p}_i$, $\mathbf{p}_j$, b, and ϵ are given, along with the intermolecular potential energy.

This derivation of the Boltzmann equation has the advantage of simplicity and direct physical interpretation. However, some aspects of this treatment present logical difficulties, because of the finite extension of the molecules and because of the finite duration of a collision. To put the Boltzmann equation on a firmer foundation, Kirkwood [39] has derived it directly from the Liouville theorem.

The Boltzmann equation derived from the Liouville theorem. The state of a gas made up of N molecules is described by the distribution function $f^{(N)}(\mathbf{r}^N, \mathbf{p}^N, t)$ in the γ space of $6N$ dimensions. According to the Liouville equation, the time variation of the distribution function $f^{(N)}$ is given by the equation

$$\frac{\partial f^{(N)}}{\partial t} + \sum_i (\nabla_{p_i} H) \cdot (\nabla_{r_i} f^{(N)}) - \sum_i (\nabla_{r_i} H) \cdot (\nabla_{p_i} f^{(N)}) = 0 \quad (26\text{-}17)$$

or

$$\frac{\partial f^{(N)}}{\partial t} + \sum_{i=1}^{N} \left\{ \frac{1}{m_i} (\mathbf{p}_i \cdot \nabla_{r_i} f^{(N)}) + [(\mathbf{F}_i + \mathbf{X}_i) \cdot \nabla_{p_i} f^{(N)}] \right\} = 0 \quad (26\text{-}18)$$

where m_i is the mass of molecule i, $\mathbf{F}_i$ is the force on molecule i due to all of the other molecules, $\mathbf{X}_i$ is the force on molecule i due to an external field, and H is the Hamiltonian.

As stated earlier in this article, the macroscopic behavior of a gas is usually described with sufficient accuracy by a distribution function of lower order. For example, the macroscopic behavior of a gas at sufficiently low densities is described by the set of distribution functions $f_i^{(1)}$. These functions are defined as the integral of $f^{(N)}$ over the coordinates and momenta of all but one of the molecules. Because of the symmetry of $f^{(N)}$, this function depends only upon the species of the remaining molecule, indicated by the subscript i, and does not depend upon which molecule of the particular kind is chosen as special.

An equation for $f_i^{(1)}$ may be obtained from the Liouville equation (26-18) by integrating over the coordinates of $(N - 1)$ molecules. When such an integration is performed and use is made of the fact that $f_i^{(1)}$ vanishes when $|\mathbf{p}_i| \to \infty$ and also at the walls of the containing vessel, one obtains

$$\frac{\partial f_i^{(1)}}{\partial t} + \frac{1}{m_i} (\mathbf{p}_i \cdot \nabla f_i^{(1)}) + (\mathbf{X}_i \cdot \nabla_{p_i} f_i^{(1)}) = -\frac{1}{(N-1)!} \iint (\mathbf{F}_i \cdot \nabla_{p_i} f^{(N)}) d\mathbf{r}^{N-1} d\mathbf{p}^{N-1} \quad (26\text{-}19)$$

This equation does not in itself define the behavior of $f_i^{(1)}$. As discussed above, whenever one lowers the level of description it is necessary to introduce a condition which effectively restricts the nature of the systems under consideration. In this case, in order to obtain the Boltzmann equation it is necessary to introduce the concept of "molecular chaos."

Kirkwood [39] showed that in the derivation of the Boltzmann equation, the assumption is made implicitly that the distribution functions, $f_i^{(1)}(t)$, do not change appreciably during the time of a collision. This may be seen in the following way: For the case of a gas containing a single component and having only two-body forces between the molecules, Eq. 26-19 reduces to

$$\frac{\partial f_1^{(1)}}{\partial t} + \frac{1}{m}(\mathbf{p}_1 \cdot \nabla f_1^{(1)}) + (\mathbf{X}_1 \cdot \nabla_{p_1} f_1^{(1)}) = -\iint (\mathbf{F}_{12} \cdot \nabla_{p_1} f_{12}^{(2)}) d\mathbf{r}_2 d\mathbf{p}_2 \quad (26\text{-}20)$$

where $\mathbf{F}_{12}$ is the force on molecule 1 due to molecule 2. If the intermolecular forces are short range, we can define a collision diameter r_0, such that $\mathbf{F}_{12}$ is effectively zero when $|\mathbf{r}_1 - \mathbf{r}_2| \geqq r_0$. All of the contributions to the integral of Eq. 26-20 come from regions where $|\mathbf{r}_1 - \mathbf{r}_2| < r_0$. The principle of molecular chaos assumes that outside this interaction sphere,

$$f_{12}^{(2)}(t) = f_1^{(1)}(t) f_2^{(1)}(t) \qquad |\mathbf{r}_1 - \mathbf{r}_2| \geqq r_0 \quad (26\text{-}21)$$

But inside of the interaction sphere, the pair distribution function, $f_{12}^{(2)}(t)$, is not known explicitly. Let us propose a scheme by which $f_{12}^{(2)}(t)$ could be calculated in principle.

If we neglect the possibility of three-body collisions, there is only one trajectory in two-particle phase space passing through the point $\mathbf{r}_1, \mathbf{p}_1; \mathbf{r}_2, \mathbf{p}_2$. Thus following the trajectory backward in time, if the system is at $\mathbf{r}_1, \mathbf{p}_1; \mathbf{r}_2, \mathbf{p}_2$ at the time t, it must have been at a well-defined point $\mathbf{r}'_1, \mathbf{r}'_2; \mathbf{p}'_1, \mathbf{p}'_2$ at time $t - \delta t$. Let us define the time $\delta t(\mathbf{r}_1, \mathbf{r}_2; \mathbf{p}_1, \mathbf{p}_2; t)$ as the interval of time such that $|\mathbf{r}'_1 - \mathbf{r}'_2| = r_0$. At this time, $f_{12}^{(2)}(\mathbf{r}'_1, \mathbf{r}'_2; \mathbf{p}'_1, \mathbf{p}'_2, t - \delta t) = f_1^{(1)}(\mathbf{r}'_1, \mathbf{p}'_1, t - \delta t) f_2^{(1)}(\mathbf{r}'_2, \mathbf{p}'_2, t - \delta t)$. Thus it follows that,

$$f_{12}^{(2)}(\mathbf{r}_1, \mathbf{r}_2; \mathbf{p}_1, \mathbf{p}_2, t) = f_1^{(1)}(\mathbf{r}'_1, \mathbf{p}'_1, t - \delta t) f_2^{(1)}(\mathbf{r}'_2, \mathbf{p}'_2, t - \delta t) \quad (26\text{-}22)$$

Thus for every point $\mathbf{r}_2, \mathbf{p}_2$ in the integral of Eq. 26-20 the pair distribution function is related to one particle distribution function at a time $t - \delta t$ and for each point there is a different δt. The magnitude of δt is of the order of the duration of a collision and is small compared with the times involved in macroscopic measurements.

Kirkwood corrected for the existence of the various δt corresponding to different points $\mathbf{r}_2, \mathbf{p}_2$ by time averaging Eq. 26-19 over an interval somewhat longer than the duration of a collision. This time-averaging is denoted by a bar so that the time-averaged distribution function is $\bar{f}_i^{(1)}$. In this manner he obtained

$$\frac{\partial \bar{f}_i^{(1)}}{\partial t} + \frac{1}{m_i}(\mathbf{p}_i \cdot \nabla \bar{f}_1^{(1)}) + (\mathbf{X}_i \cdot \nabla_{p_1} \bar{f}_1^{(1)})$$

$$= 2\pi \sum_j \iiint (\overline{f_i^{(1)\prime} f_j^{(1)\prime}} - \overline{f_i^{(1)} f_j^{(1)}}) g_{ij} b\, db\, d\epsilon\, d\mathbf{p}_j \quad (26\text{-}23)$$

This equation would be the same as the simple Boltzmann equation derived in the previous subarticle if $\overline{f_i^{(1)}f_j^{(1)}} = \bar{f}_i^{(1)}\bar{f}_j^{(1)}$. This condition is satisfied provided the distribution functions do not change appreciably in the interval of time τ_0 (comparable to the duration of a collision) over which they are time-averaged. This may be seen in the following way:

$$\overline{f_i^{(1)}f_j^{(1)}} = \frac{1}{\tau_0}\int_{-\tau_0/2}^{\tau_0/2} f_i(t+\tau)f_j(t+\tau)d\tau$$

$$= f_i^{(1)}(t)f_j^{(1)}(t) + \frac{\tau_0^2}{24}\left(f_i^{(1)}\frac{\partial^2 f_j^{(1)}}{\partial t^2} + 2\frac{\partial f_i^{(1)}}{\partial t}\frac{\partial f_j^{(1)}}{\partial t} + f_j^{(1)}\frac{\partial^2 f_i^{(1)}}{\partial t^2}\right) \quad (26\text{-}24)$$

$$\bar{f}_i^{(1)}\bar{f}_j^{(1)} = \left(\frac{1}{\tau_0}\int_{-\tau_0/2}^{\tau_0/2} f_i^{(1)}(t+\tau)d\tau\right)\left(\frac{1}{\tau_0}\int_{-\tau_0/2}^{\tau_0/2} f_j^{(1)}(t+\tau)d\tau\right)$$

$$= f_i^{(1)}(t)f_j^{(1)}(t) + \frac{\tau_0^2}{24}\left(f_i^{(1)}\frac{\partial^2 f_j^{(1)}}{\partial t^2} + f_j^{(1)}\frac{\partial^2 f_i^{(1)}}{\partial t^2}\right)$$

In the usual derivations of the Boltzmann equation δt is neglected, and it is argued that the error introduced thereby is comparable to the error introduced by neglecting in the distribution function the distance between the centers of the molecules on collision. Both errors should be negligible provided the distribution functions do not change appreciably in times of the order of the collision duration nor in distances of the order of a molecular diameter.

The distribution in velocities. In most kinetic theory problems it is more convenient to work in terms of velocities rather than momenta. Hence in the remainder of this chapter we shall use the distribution function in coordinate-velocity space $f_i(\mathbf{r}, \mathbf{v}_i, t)$, rather than the distribution function in coordinate momentum space $f_i^{(1)}(\mathbf{r}, \mathbf{p}_1, t)$. In terms of this function the Boltzmann equation is[34]

$$\frac{\partial f_i}{\partial t} + (\mathbf{v}_i \cdot \nabla f_i) + \frac{1}{m_i}(\mathbf{X}_i \cdot \nabla_{v_i} f_i) = 2\pi \sum_j \iiint (f_i' f_j' - f_i f_j) g_{ij} b\, db\, d\epsilon\, d\mathbf{v}_j \quad (26\text{-}25)$$

This equation forms the basis for the discussion of the transport properties of gases.

B,27. The Equations of Change. The hydrodynamic equations of change, the equations of conservation of mass, momentum, and energy, may be derived directly from the Boltzmann equation. Certain expressions involving the distribution functions appear in the derivation of these equations. These expressions may be shown to represent the flux of mass,

[34] We omit the superscript (1) when the distribution function is written in terms of the velocities rather than the momenta.

momentum, and energy and are directly related to the diffusion velocity, the pressure tensor, and the heat flux. These relations are derived in the present article. The approximate solutions of the Boltzmann equation and evaluation of the fluxes are discussed in subsequent articles. We begin this article by presenting a set of definitions of the various velocities which are needed to discuss the hydrodynamic equations.

Molecular velocities and stream velocities. The *linear velocity* of a molecule of species j with respect to a coordinate system fixed in space is denoted by $\mathbf{v}_j$, with components v_{jx}, v_{jy}, v_{jz}. Its magnitude $|\mathbf{v}_j|$, or simply v_j, is called the molecular speed. For a chemical species j present at a number density n_j we define an *average velocity* $\bar{\mathbf{v}}_j$ by

$$\bar{\mathbf{v}}_j(\mathbf{r}, t) = \frac{1}{n_j}\int \mathbf{v}_j f_j(\mathbf{r}, \mathbf{v}_j, t)d\mathbf{v}_j \tag{27-1}$$

The average velocity is a function of position and time, and represents the macroscopic rate of flow of the chemical species j. The bar denotes, in general, the average value of a function of the velocity, e.g.

$$\bar{\alpha}(\mathbf{r}, t) = \frac{1}{n_j}\int \alpha(\mathbf{v}_j) f_j(\mathbf{r}, \mathbf{v}_j, t)d\mathbf{v}_j \tag{27-2}$$

is the average value of $\alpha(\mathbf{v}_j)$.

The *mass average velocity* is defined by

$$\mathbf{v}_0(\mathbf{r}, t) = \frac{1}{\rho}\sum n_j m_j \bar{\mathbf{v}}_j \tag{27-3}$$

in which $\rho(\mathbf{r}, t)$ is the over-all density of the gas at a particular point

$$\rho(\mathbf{r}, t) = \sum n_j m_j \tag{27-4}$$

The mass average velocity is then a weighted mean, with each molecule being given a weight proportional to its mass. The momentum of the gas per unit volume is the same as if all the molecules were moving with the mass average velocity $\mathbf{v}_0$. This velocity is also referred to as the *stream velocity* or the *flow velocity*.

The *peculiar velocity* of a molecule of species j is defined as the velocity of the molecule with respect to an axis moving with the mass average velocity $\mathbf{v}_0$

$$\mathbf{V}_j(\mathbf{v}_j, \mathbf{r}, t) = \mathbf{v}_j - \mathbf{v}_0 \tag{27-5}$$

The *diffusion velocity* of chemical species j is the rate of flow of molecules of j with respect to the mass average velocity of the gas,

$$\bar{\mathbf{V}}_j(\mathbf{r}, t) = \bar{\mathbf{v}}_j - \mathbf{v}_0 \tag{27-6}$$

Clearly, the diffusion velocity is also the average of the peculiar velocity and may be written in the form

$$\bar{\mathbf{V}}_j(\mathbf{r}, t) = \frac{1}{n_j} \int (\mathbf{v}_j - \mathbf{v}_0) f_j(\mathbf{r}, \mathbf{v}_j, t) d\mathbf{v}_j \tag{27-7}$$

as a consequence of the definitions of $\bar{\mathbf{v}}_j$ and $\mathbf{V}_j$. (Note that $\mathbf{v}_0$ is a function of position and time, but not of the velocities $\mathbf{v}_j$.) It follows from the definitions of the diffusion velocity and the mass average velocity that

$$\sum n_j m_j \bar{\mathbf{V}}_j = \sum n_j m_j (\bar{\mathbf{v}}_j - \mathbf{v}_0) = 0 \tag{27-8}$$

For the sake of simplicity in writing certain expressions in subsequent articles, a *reduced velocity* $\mathbf{W}_j$ is used frequently. It is defined by

$$\mathbf{W}_j = \sqrt{\frac{m_j}{2kT}}\, \mathbf{V}_j \tag{27-9}$$

and its magnitude is designated by W_j.

We define the *kinetic theory temperature* in terms of the mean peculiar kinetic energy, averaged over all types of molecules

$$\frac{3}{2} kT = \frac{1}{n} \sum n_j \left(\frac{1}{2} m_j \overline{V_j^2}\right) \tag{27-10}$$

where

$$n = \sum n_j \tag{27-11}$$

is the total number density of the mixture. The concept of temperature was introduced in Art. 24 through thermodynamic arguments. It is of course necessary that the present definition be consistent with the earlier one. Since, however, the earlier discussion was restricted to equilibrium conditions, the previous definition has meaning only under these conditions. That the present more general definition reduces correctly in this limit is evident from the fact that the equilibrium distribution function (the Maxwellian distribution, given by Eq. 28-7) is a special case of the Boltzmann distribution function of Art. 25.

The flux vectors. In a gas under nonequilibrium conditions, gradients exist in one or more of the macroscopic physical properties of the system: composition, mass average velocity, and temperature. The gradients of these properties are the cause of the molecular transport of mass (m_j), momentum ($m_j\mathbf{V}_j$), and kinetic energy ($\frac{1}{2}m_j\mathbf{V}_j^2$) through the gas. Since the mechanism of the transport of each of these molecular properties can be treated similarly, they are designated collectively by ψ_j.

Let us now examine the transport of these properties ψ_j on the microscopic level. Imagine in the gas a small element of surface, dS, moving with the mass average velocity $\mathbf{v}_0$. The orientation of the surface is designated by a unit vector $\mathbf{n}$ normal to the surface. Then the velocity of molecules of the jth species with respect to the surface element dS is $\mathbf{V}_j$, according to the definition in Eq. 27-5. All those molecules of the jth species which have velocity[35] $\mathbf{V}_j$ and which cross dS during a time interval dt must at the beginning of this time interval be located in a cylinder with dS as its base and with generators parallel to $\mathbf{V}_j$ and of length $|\mathbf{V}_j|dt$ (see Fig. B,27a). This cylinder has a volume $(\mathbf{n}dS \cdot \mathbf{V}_j dt)$.

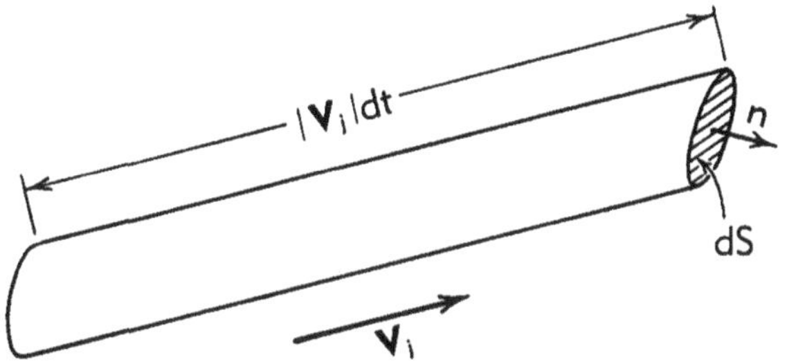

Fig. B,27a. Elemental cylinder.

Since there are $f_j d\mathbf{V}_j$ molecules per unit volume which have a velocity $\mathbf{V}_j$, the number of molecules which cross dS during a time interval dt is

$$(\mathbf{n} \cdot \mathbf{V}_j) f_j d\mathbf{V}_j dS dt \tag{27-12}$$

If associated with each molecule there is some property ψ_j, the magnitude of which depends on $\mathbf{V}_j$, then

$$(\mathbf{n} \cdot \mathbf{V}_j) \psi_j f_j d\mathbf{V}_j dS dt \tag{27-13}$$

is the amount of this property transported across dS during the time interval dt by molecules with velocities in the range $d\mathbf{V}_j$ about $\mathbf{V}_j$. The flux of this property (that is, the amount which crosses per unit area per unit time) is then

$$\psi_j f_j (\mathbf{n} \cdot \mathbf{V}_j) d\mathbf{V}_j \tag{27-14}$$

The total flux across the elementary surface is obtained by adding the contributions from molecules within all velocity ranges, and is accordingly:

$$\int \psi_j f_j (\mathbf{n}_j \cdot \mathbf{V}_j) d\mathbf{V}_f = (\mathbf{n} \cdot \quad \psi_j f_j \mathbf{V}_j d\mathbf{V}_j) \equiv (\mathbf{n} \cdot \boldsymbol{\Psi}_j) \tag{27-15}$$

The vector

$$\boldsymbol{\Psi}_j = \int \psi_j f_j \mathbf{V}_j d\mathbf{V}_j \tag{27-16}$$

is called the "flux vector" associated with the property ψ_j. This vector has the physical significance that the component of the vector in any direction $\mathbf{n}$ is the flux of the associated physical property across a surface normal to $\mathbf{n}$.

[35] Or more precisely those molecules whose velocities lie in a small range $d\mathbf{V}_j$ about $\mathbf{V}_j$.

Let us now examine the flux vectors related to the transport of mass, momentum, and kinetic energy:

1. Transport of Mass

   If $\psi_j = m_j$, then[36]

$$\boldsymbol{\Psi}_j = m_j \int f_j \mathbf{V}_j d\mathbf{V}_j = n_j m_j \overline{\mathbf{V}}_j \tag{27-17}$$

is the flux vector associated with the transport of mass.

2. Transport of Momentum

   If $\psi_j = m_j(\mathbf{v}_j - \mathbf{v}_0)_x = m_j V_{jx}$, then

$$\boldsymbol{\Psi}_j = m_j \int V_{jx} \mathbf{V}_j f_j d\mathbf{V}_j = n_j m_j \overline{V_{jx} \mathbf{V}_j} \tag{27-18}$$

is the flux vector associated with the transport of the x component of momentum (relative to $\mathbf{v}_0$). This vector has components proportional to $\overline{V_{jx}V_{jx}}$, $\overline{V_{jx}V_{jy}}$, and $\overline{V_{jx}V_{jz}}$. Similar flux vectors can be obtained for the y and z components of the momentum, making a total of three flux vectors associated with momentum transfer. The nine components of these three vectors form a symmetric second order tensor, p,

$$\begin{aligned} (p_j)_{xx} &= m_j \int f_j V_{jx} V_{jx} d\mathbf{V}_j = n_j m_j \overline{V_{jx} V_{jx}} \\ (p_j)_{xy} &= (p_j)_{yx} = n_j m_j \overline{V_{jx} V_{jy}} \\ &\cdots \end{aligned} \tag{27-19}$$

Symbolically,

$$\mathsf{p}_j = n_j m_j \overline{\mathbf{V}_j \mathbf{V}_j} \tag{27-20}$$

is the tensor associated with the partial pressure of the jth chemical species in the gas. The sum of the partial pressure tensors over all the species in the gas gives the pressure tensor for the mixture:

$$\mathsf{p} = \sum \mathsf{p}_j = \sum n_j m_j \overline{\mathbf{V}_j \mathbf{V}_j} \tag{27-21}$$

The pressure tensor has the physical significance that it represents the flux of momentum through the gas. The individual components have the following meaning: The diagonal elements, p_{xx}, p_{yy}, p_{zz} are *normal stresses;* that is, p_{xx} is the force per unit area in the x direction exerted on a plane surface in the gas which is perpendicular to the x direction. The nondiagonal elements are *shear stresses;* that is, p_{yx} represents the force per unit area in the x direction exerted on a plane surface which is perpendicular to the y direction. The significance of the components, p_{yx}, p_{yy}, p_{yz}, which when combined together give a resultant force, $\mathbf{p}_y$, on a unit area perpendicular to the y direction, is shown in Fig. B,27b. The pressure tensor represents stresses or pressures measured by an instru-

[36] It is to be noted that in computing average quantities, integration over V_j is equivalent to integration over $\mathbf{v}_j$ since the two differ by a constant and the integration is over the entire range.

ment moving with the stream velocity $\mathbf{v}_0$. The pressure as measured by a stationary gauge depends upon $\mathbf{v}_0$ and the orientation of the gauge.

It is shown in Art. 29 that at equilibrium the shear stresses are zero and the normal stresses are equal. In this case the force on any surface element in the gas is constant and normal to the surface regardless of its orientation; that is

$$p_{xx} = p_{yy} = p_{zz} = p \tag{27-22}$$

where p is the equilibrium hydrostatic pressure.

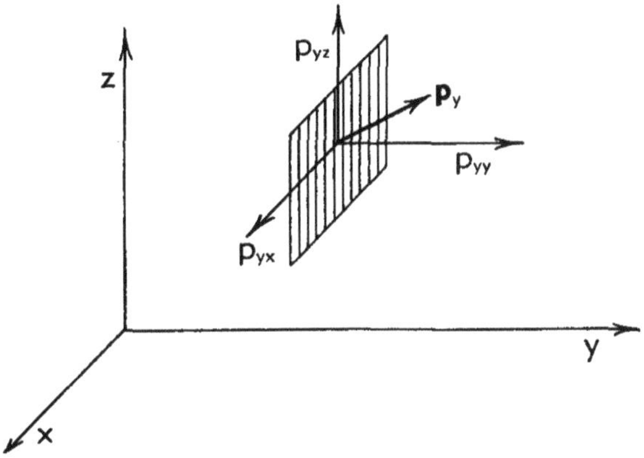

Fig. B,27b. Significance of the components of the pressure tensor. In this figure is shown a small element of surface whose normal is in the y direction. On this element, located in the body of a system of gas, there is a total force per unit area, $\mathbf{p}_y$. The components of this force are p_{yy} (a normal stress) and p_{yz} and p_{yx} (shear stresses). The nine components of this type make up the "pressure tensor."

3. Transport of Kinetic Energy

If $\psi_j = \frac{1}{2}m_j(\mathbf{v}_j - \mathbf{v}_0)^2 = \frac{1}{2}m_jV_j^2$, then

$$\mathbf{\Psi}_j = \tfrac{1}{2}m_j \int V_j^2\mathbf{V}_jf_jd\mathbf{V}_j = \tfrac{1}{2}n_jm_j\overline{V_j^2\mathbf{V}_j} \tag{27-23}$$

is the flux vector associated with the transport of kinetic energy for molecules of the jth species. The sum of such vectors over all the components in the gas mixture is the "heat flux vector" $\mathbf{q}$:

$$\mathbf{q} = \sum \tfrac{1}{2}n_jm_j\overline{V_j^2\mathbf{V}_j} \tag{27-24}$$

This vector has the physical significance that its components q_x, q_y, and q_z represent the flux of kinetic energy in the x, y, and z directions respectively.

It should be noted that these are the correct expressions for the flux vectors in a dilute gas only. In a more dense gas there are additional terms due to "collisional transfer." When two molecules undergo a collision, a certain amount of momentum and energy is transferred almost instantaneously from the center of one molecule to that of the other. Now if in Fig. B,27a the centers of two molecules should be on the opposite sides of the surface element dS, the resulting collisional transfer provides a contribution to the fluxes which is not included in the above expressions.[37] This limitation will not be of importance in this chapter inasmuch as the effect of collisional transfer is negligible under conditions for which the Boltzmann equation which we use is valid.

The general equation of change.[38] The fundamental hydrodynamic equations of continuity, motion, and energy balance may be derived from the Boltzmann equation without actually determining the form of the distribution function f_i. If the Boltzmann equation (Eq. 26-25) for the ith component is multiplied by the quantity ψ_i associated with the ith species, and if the equation is integrated over $\mathbf{v}_i$, one obtains

$$\int \psi_i \left[\frac{\partial f_i}{\partial t} + (\mathbf{v}_i \cdot \nabla f_i) + \frac{1}{m_i}(\mathbf{X}_i \cdot \nabla_{v_i} f_i)\right] d\mathbf{v}_i$$

$$= \sum_j \iiiint \psi_i(f_i' f_j' - f_i f_j) g_{ij} b\,db\,d\epsilon\,d\mathbf{v}_i d\mathbf{v}_j \quad (27\text{-}25)$$

The three terms on the left-hand side of Eq. 27-25 may be transformed by simple manipulations:

$$\int \psi_i \frac{\partial f_i}{\partial t} d\mathbf{v}_i = \frac{\partial}{\partial t}\int \psi_i f_i d\mathbf{v}_i - \int f_i \frac{\partial \psi_i}{\partial t} d\mathbf{v}_i = \frac{\partial(n_i \overline{\psi}_i)}{\partial t} - n_i \overline{\frac{\partial \psi_i}{\partial t}} \quad (27\text{-}26)$$

$$\int \psi_i v_{ix} \frac{\partial f_i}{\partial x} d\mathbf{v}_i = \frac{\partial}{\partial x}\int \psi_i v_{ix} f_i d\mathbf{v}_i - \int f_i v_{ix} \frac{\partial \psi_i}{\partial x} d\mathbf{v}_i$$

$$= \frac{\partial}{\partial x}(n_i \overline{\psi_i v_{ix}}) - n_i \overline{v_{ix} \frac{\partial \psi_i}{\partial x}} \quad (27\text{-}27)$$

$$\int \psi_i \frac{\partial f_i}{\partial v_{ix}} d\mathbf{v}_i = \iint [\psi_i f_i]_{v_{ix}=-\infty}^{v_{ix}=+\infty} dv_{iy} dv_{iz} - \int f_i \frac{\partial \psi_i}{\partial v_{ix}} d\mathbf{v}_i$$

$$= -n_i \overline{\frac{\partial \psi_i}{\partial v_{ix}}} \quad (27\text{-}28)$$

The last two equations have been written for only the x components.[39] In Eq. 27-28 the first term produced by the partial integration vanishes because the product $f_i\psi_i$ is assumed to diminish sufficiently rapidly for

[37] A detailed discussion of this has been given by Kirkwood [*40*]. Collisional transfer is very important in the theory of dense gases and liquids; see [*34*, Chap. 9].

[38] Although the derivation given here is valid only for dilute gases, the forms of the equations of change as derived in this article are valid even for dense gases and liquids.

[39] The quantity ψ_i may depend on $\mathbf{r}$ and t through $\mathbf{v}_0(\mathbf{r}, t)$. The quantities $\mathbf{r}$, $\mathbf{v}_i$, and t are taken as independent variables.

large v_i. We may use these last three equations to obtain

$$\frac{\partial n_i\overline{\psi_i}}{\partial t} + (\nabla \cdot n_i\overline{\psi_i\mathbf{v}_i}) - n_i\left[\overline{\frac{\partial \psi_i}{\partial t}} + (\overline{\mathbf{v}_i \cdot \nabla\psi_i}) + \left(\frac{\mathbf{X}_i}{m_i} \cdot \overline{\nabla_{v_i}\psi_i}\right)\right]$$
$$= \sum_j \iiiint \psi_i(f_i'f_j' - f_if_j)g_{ij}b\,db\,d\epsilon\,d\mathbf{v}_i d\mathbf{v}_j \quad (27\text{-}29)$$

This is known as Enskog's general "equation of change" for a physical quantity ψ_i associated with the ith kind of molecule. Summation over i gives the equation of change for the property ψ for the entire gas.

The vanishing of the collision integrals for the summational invariants. The equations of the type shown in Eq. 27-29 are not particularly useful for a general ψ_i because of the very complex integrals which occur on the right side. However, if ψ_i is the mass of the ith molecule, then because the masses of the individual molecules are unchanged in a collision,

$$\iiiint m_i(f_i'f_j' - f_if_j)g_{ij}b\,db\,d\epsilon\,d\mathbf{v}_i d\mathbf{v}_j = 0 \quad (27\text{-}30)$$

Since mass, momentum, and energy are conserved during a collision, it follows that if ψ_i is m_i, $m_i\mathbf{V}_i$, or $\frac{1}{2}m_iV_i^2$ then

$$\sum_{i,j} \iiiint \psi_i(f_i'f_j' - f_if_j)g_{ij}b\,db\,d\epsilon\,d\mathbf{v}_i d\mathbf{v}_j = 0 \quad (27\text{-}31)$$

These equations enable one to simplify the general equation of change.

The validity of these last two equations may be justified by the following arguments: The integral,

$$\iiiint \psi_i(f_i'f_j' - f_if_j)g_{ij}b\,db\,d\epsilon\,d\mathbf{v}_i d\mathbf{v}_j \quad (27\text{-}32)$$

is equal to the integral

$$\iiiint \psi_i'(f_if_j - f_i'f_j')g_{ij}'b'\,db'\,d\epsilon\,d\mathbf{v}_i' d\mathbf{v}_j' \quad (27\text{-}33)$$

which is written in terms of inverse encounters, i.e. different symbols are used to indicate the integration variables. In Art. 26 it is stated that

$$g_{ij} = g_{ij}', \qquad b = b', \quad \text{and} \quad d\mathbf{v}_i d\mathbf{v}_j = d\mathbf{v}_i' d\mathbf{v}_j' \quad (27\text{-}34)$$

so that Eq. 27-33 may be rewritten as

$$-\iiiint \psi_i'(f_i'f_j' - f_if_j)g_{ij}b\,db\,d\epsilon\,d\mathbf{v}_i d\mathbf{v}_j \quad (27\text{-}35)$$

Since the integrals of Eq. 27-32 and 27-35 are equal, they are also each equal to one-half the sum of the two. The result of this "symmetrizing" operation is that

$$\iiiint \psi_i(f_i'f_j' - f_if_j)g_{ij}b\,db\,d\epsilon\,d\mathbf{v}_i d\mathbf{v}_j$$
$$= \frac{1}{2}\iiiint (\psi_i - \psi_i')(f_i'f_j' - f_if_j)g_{ij}b\,db\,d\epsilon\,d\mathbf{v}_i d\mathbf{v}_j \quad (27\text{-}36)$$

For the case that $\psi_i = m_i$ this equation shows immediately the validity of Eq. 27-30; for $(\psi_i - \psi_i') = 0$ expresses the fact that the masses of the individual molecules are not changed during an encounter.

Eq. 27-36 may be summed over both i and j; then the dummy indices may be interchanged to give the two equivalent expressions

$$\frac{1}{2}\sum_{i,j}\iiiint (\psi_i - \psi_i')(f_i'f_j' - f_if_j)g_{ij}b\,db\,d\epsilon\,d\mathbf{v}_i\,d\mathbf{v}_j$$
$$= \frac{1}{2}\sum_{i,j}\iiiint (\psi_j - \psi_j')(f_i'f_j' - f_if_j)g_{ij}b\,db\,d\epsilon\,d\mathbf{v}_i\,d\mathbf{v}_j \quad (27\text{-}37)$$

so that

$$\sum_{i,j}\iiiint \psi_i(f_i'f_j' - f_if_j)g_{ij}b\,db\,d\epsilon\,d\mathbf{v}_i\,d\mathbf{v}_j$$
$$= \frac{1}{4}\sum_{i,j}\iiiint (\psi_i + \psi_j - \psi_i' - \psi_j')(f_i'f_j' - f_if_j)g_{ij}b\,db\,d\epsilon\,d\mathbf{v}_i\,d\mathbf{v}_j \quad (27\text{-}38)$$

The vanishing of the expression $(\psi_i + \psi_j - \psi_i' - \psi_j')$ is used to define the invariants of the encounter. Consequently, Eq. 27-31 is valid for the properties m_i, $m_i\mathbf{V}_i$, and $\frac{1}{2}m_iV_i^2$.

Explicit expressions for the equations of change. Now that the validity of Eq. 27-30 and 27-31 has been demonstrated, this information may be used to derive the equations of change for specific molecular properties. This is done by letting ψ_i in Eq. 27-29 be m_i, $m_i\mathbf{V}_i$, and $\frac{1}{2}m_iV_i^2$ and then summing over the index i.

When $\psi_i = m_i$ we may follow this procedure and use the vanishing of the collision integrals just established to obtain

$$\frac{\partial n_i}{\partial t} + (\nabla \cdot n_i\bar{\mathbf{v}}_i) = 0 \quad (27\text{-}39)$$

which, in terms of the diffusion velocity, becomes

$$\frac{\partial n_i}{\partial t} + (\nabla \cdot n_i(\mathbf{v}_0 + \overline{\mathbf{V}}_i)) = 0 \quad (27\text{-}40)$$

These last two equations represent two forms of the "equation of continuity for the ith chemical species." If in the last expression each of the equations is multiplied by m_i and the equations added, then (since $\sum n_im_i\overline{\mathbf{V}}_i = 0$)

$$\frac{\partial \rho}{\partial t} + (\nabla \cdot \rho\mathbf{v}_0) = 0 \quad (27\text{-}41)$$

This is the *equation of continuity* for the gas as a whole.

If we put $\psi_i = m_i\mathbf{V}_i$ in Eq. 27-29 and sum over i, the collision integrals on the right side of the equation vanish so that we obtain

$$\sum m_i \left[\frac{\partial n_i \overline{\mathbf{V}}_i}{\partial t} + (\nabla \cdot n_i \overline{\mathbf{v}_i \mathbf{V}_i}) - n_i \frac{\partial \overline{\mathbf{V}}_i}{\partial t} - n_i(\overline{\mathbf{v}_i \cdot \nabla \mathbf{V}_i}) - \frac{n_i}{m_i} (\mathbf{X}_i \cdot \overline{\nabla_{v_i} \mathbf{V}_i}) \right] = 0 \tag{27-42}$$

This equation may be simplified by using the relation among the diffusion velocities (Eq. 27-8), the definition of the pressure tensor (Eq. 27-21), and the definition of the diffusion velocities (Eq. 27-6). It should be noted that in the differentiations $\mathbf{r}$ and $\mathbf{v}_i$ are considered as the independent variables. Then one obtains for the *equation of motion of the gas*

$$\frac{\partial \mathbf{v}_0}{\partial t} + (\mathbf{v}_0 \cdot \nabla \mathbf{v}_0) = -\frac{1}{\rho} (\nabla \cdot \mathsf{p}) + \frac{1}{\rho} \sum_i n_i \mathbf{X}_i \tag{27-43}$$

If $\psi_i = \frac{1}{2} m_i V_i^2$ is used in the general equation of change, we may perform the manipulations just described to obtain

$$\sum \frac{1}{2} m_i \left[\frac{\partial n_i \overline{V_i^2}}{\partial t} + (\nabla \cdot n_i \overline{V_i^2 \mathbf{v}_i}) - n_i \frac{\partial \overline{V_i^2}}{\partial t} \right.$$
$$\left. - n_i(\overline{\mathbf{v}_i \cdot \nabla V_i^2}) - \left(\frac{\mathbf{X}_i}{m_i} \cdot \overline{\nabla_{v_i} V_i^2} \right) \right] = 0 \tag{27-44}$$

This equation may be transformed using the same methods as in the treatment of the equation of motion. If one introduces the pressure tensor, defined by Eq. 27-21, and the heat flux vector, defined by Eq. 27-24, one obtains the *equation of energy balance*

$$\frac{\partial}{\partial t} (\rho \hat{U}^{(\mathrm{tr})}) + (\nabla \cdot \rho \hat{U}^{(\mathrm{tr})} \mathbf{v}_0) + (\nabla \cdot \mathbf{q}) + (\mathsf{p} : \nabla \mathbf{v}_0)$$
$$- \sum n_i (\mathbf{X}_i \cdot \overline{\mathbf{V}}_i) = 0 \tag{27-45}$$

in which the notation $(\mathsf{p} : \nabla \mathbf{v}_0)$ means $\sum_i \sum_j p_{ij} (\partial V_{0i} / \partial x_j)$, and $\hat{U}^{(\mathrm{tr})}$ is the translational contribution to the internal energy per unit mass defined by:[40]

$$\hat{U}^{(\mathrm{tr})} = \frac{1}{\rho} \sum \frac{1}{2} n_i m_i \overline{V_i^2} \tag{27-46}$$

This quantity represents the total energy of unit mass of gas in a coordinate frame moving with the mass average velocity $\mathbf{v}_0$ (that is, the kinetic energy of the over-all flow is not included), excluding the potential energy due to the external field.[41] If one uses the equation of continuity

[40] For molecules without internal degrees of freedom this is the only contribution to the internal energy. In general, however, there are other contributions.

[41] Some authors have written the energy balance equation in terms of an energy which includes either the kinetic energy associated with $\mathbf{v}_0$ or the potential energy associated with the external force, or both.

(Eq. 27-41), the equation of energy balance (Eq. 27-45) may be written in a somewhat different form:

$$\rho \frac{\partial \hat{U}^{(tr)}}{\partial t} + \rho(\mathbf{v}_0 \cdot \nabla \hat{U}^{(tr)}) = -(\nabla \cdot \mathbf{q}) - (\mathsf{p}:\nabla \mathbf{v}_0) + \sum n_i(\mathbf{X}_i \cdot \overline{\mathbf{V}}_i) \quad (27\text{-}47)$$

This may be restated in terms of the temperature by using the definitions of $\hat{U}^{(tr)}$ and T

$$\frac{3}{2} nk \left[\frac{\partial T}{\partial t} + (\mathbf{v}_0 \cdot \nabla T)\right] = -(\nabla \cdot \mathbf{q}) - (\mathsf{p}:\nabla \mathbf{v}_0) + \sum n_i(\mathbf{X}_i \cdot \overline{\mathbf{V}}_i) + \frac{3}{2} kT \left(\nabla \cdot \sum n_i \overline{\mathbf{V}}_i\right) \quad (27\text{-}48)$$

These last two equations are two forms of the energy balance equation. It may be shown that Eq. 27-47 is quite general and applies even in the case of a reacting mixture of polyatomic molecules, if $\hat{U}^{(tr)}$ is replaced by total internal energy per gram. However, Eq. 27-48 applies only to a non-reacting mixture of particles which have no internal degrees of freedom.

B,28. Enskog's Solution of the Boltzmann Equation. Various attempts have been made to obtain approximate solutions to the Boltzmann equation. We consider here the perturbation technique of Enskog, which is a modification of a method due to Hilbert. This method of successive approximations can, in principle, be extended to systems in which the gradients of the thermodynamic quantities are quite large. In the zeroeth approximation, the distribution function is locally Maxwellian, and one obtains the Eulerian equations of change. The first order perturbation leads to the Navier-Stokes[42] equations; the second order perturbation gives the Burnett equations. From the higher approximations there result more complicated equations in which the flux vectors depend progressively on higher derivatives of the thermodynamic quantities and higher powers of the lower derivatives. The results of the higher order perturbations are seldom used.

A gas in any initial state which is permitted to remain undisturbed for a sufficient length of time approaches a stationary state. If the gas is isolated adiabatically and not subject to an external force, the stationary state is a uniform condition in which all of the distribution functions are Maxwellian. The proof that the equilibrium distribution functions are Maxwellian (the H-theorem) is discussed below. The remaining portion of the article is devoted to a discussion of Enskog's solution of the Boltzmann equation, and includes a detailed consideration of the results of

[42] The Navier-Stokes equations apply to systems in which the gradients in the physical properties are small, i.e. in which the physical properties do not change appreciably in a distance of mean free path.

the first order perturbation. This method involves the solution of a set of integral equations by a variational procedure which leads to a rapidly converging series.

The H-theorem. The distribution functions describing the behavior of a gaseous mixture are the solutions of the set of Boltzmann equations, one for each component. From these equations, which are derived in Art. 26 and are given in final form in Eq. 26-25, we find that in the absence of external forces and under uniform conditions the time rate of change of the distribution functions f_i is given by

$$\frac{\partial f_i}{\partial t} = \sum_j \iiint (f_i'f_j' - f_if_j)g_{ij}bdbd\epsilon d\mathbf{v}_j \tag{28-1}$$

That these functions approach a limiting form which is the Maxwellian distribution function is evident from the treatment of equilibrium statistical mechanics of Chap. 4. It is, however, of interest to show that the Boltzmann equations lead to an identical set of equilibrium distribution functions.

In order to examine the equilibrium solution it is convenient to introduce a function $H(t)$ defined by

$$H(t) = \sum_i \int f_i(\mathbf{v}_i, t) \ln f_i(\mathbf{v}_i, t) d\mathbf{v}_i \tag{28-2}$$

This function is a generalization of the entropy defined in Chap. 4 and the proof of the H-theorem given here is a special case of the general statistical proof of the second law of thermodynamics. Differentiating $H(t)$ and making use of the Boltzmann equation for $\partial f_i/\partial t$, we obtain

$$\frac{dH}{dt} = 2\pi \sum_{i,j} \iiiint (1 + \ln f_i)(f_i'f_j' - f_if_j)g_{ij}bdbd\epsilon d\mathbf{v}_i d\mathbf{v}_j \tag{28-3}$$

The integral on the right-hand side of this equation may be symmetrized by the method explained in the previous article to give

$$\frac{dH}{dt} = -\frac{1}{4} \sum_{i,j} \iiiint \left[\ln \frac{f_i'f_j'}{f_if_j}\right] (f_i'f_j' - f_if_j)g_{ij}bdbd\epsilon d\mathbf{v}_i d\mathbf{v}_j \tag{28-4}$$

Each of the integrands is of the form $(x - y) \ln (x/y)$ where in each case $x = f_i'f_j'$ and $y = f_if_j$. If $x > y$ then both $(x - y)$ and $\ln (x/y)$ are positive; if $x < y$ then both $(x - y)$ and $\ln (x/y)$ are negative. Therefore the integrand of each of the integrals on the right of this equation is always positive or zero. Hence dH/dt is negative or zero so that $H(t)$ can never increase. From the definition of H it follows that it is bounded, and thus approaches a limit[43] for large values of t. In this limit the distribution functions are such that integrands of each of the integrals on the right of Eq. 28-4 are identically zero. Thus, at equilibrium,

$$\ln f_i + \ln f_j = \ln f_i' + \ln f_j' \tag{28-5}$$

which states that the logarithms of the distribution functions are summational invariants of a molecular collision.

It can be shown that the only summational invariants are linear combinations of the three (five scalar) invariants: the mass m_i, the momentum $m_i\mathbf{v}_i$, and the kinetic energy $\frac{1}{2}m_iv_i^2$. Thus at equilibrium the most general expression for $\ln f_i$ is of the form

$$\ln f_i = a_im_i + [\mathbf{b}_i \cdot (m_i\mathbf{v}_i)] + c_i(\tfrac{1}{2}m_iv_i^2) \tag{28-6}$$

where a_i, $\mathbf{b}_i$, and c_i are constants depending (through the initial distribution functions) on the total number of molecules of kind i, the total momentum, and the total energy of the system. It is convenient to write this expression in terms of the physical parameters: n_i, the number density of molecules of kind i; $\mathbf{v}_0$, the mass average velocity, defined by Eq. 27-3; and the temperature T, defined by Eq. 27-10. In terms of these quantities the equilibrium distribution is

$$f_i = n_i\left(\frac{m_i}{2\pi kT}\right)^{\frac{3}{2}} e^{-m_iV_i^2/2kT} \tag{28-7}$$

where $\mathbf{V}_i = (\mathbf{v}_i - \mathbf{v}_0)$ is the peculiar velocity defined in Eq. 27-5.

At equilibrium, f_i is independent of time and hence the term on the right of Eq. 28-1 is zero. This constitutes a state of over-all balance in the collision processes: The number of molecules of kind i in a particular velocity range which are lost due to collisions is exactly compensated by the number created by the collision processes. However, the H-theorem states that the only equilibrium solution[44] to the Boltzmann equations is one in which not only the term on the right of Eq. 28-1 is zero but also the integrand of each integral is identically zero. This is proof of the statistical principle of "detailed balance." That is, at equilibrium the number of molecules of kind i in a particular velocity range which are lost due to collisions of a particular kind with molecules of kind j is exactly balanced by the number of reverse collisions.

[43] For simplicity consider a single pure component rather than a mixture. The function H can equal $-\infty$ only if the integral, $(a) = \int f \ln f d\mathbf{v}$ fails to converge. Now, in any case the integral, $(b) = \int \frac{1}{2}mv^2 f d\mathbf{v}$ which corresponds to the total kinetic energy of the molecules, must converge. A comparison of the integrals (a) and (b) indicates that (a) must converge provided that $\ln f$ does not approach infinity faster than v^2. However, if $\ln f$ approached infinity faster than v^2 it would imply that f itself decreased faster than $e^{-\frac{1}{2}(mv^2)}$ in which case the integral (a) would surely converge. Chapman and Cowling [38, p. 70] consider this problem carefully.

[44] The proof of this rests basically on the proof that the only summational invariants are linear combinations of the five scalars mentioned above.

The Enskog series [41]. The Boltzmann equations (26-25) may be written as

$$\frac{\partial f_i}{\partial t} + (\mathbf{v}_i \cdot \nabla f_i) + \frac{1}{m_i}(\mathbf{X}_i \cdot \nabla_{v_i} f_i) = \sum_j J(f_i, f_j) \tag{28-8}$$

where $J(f_i, f_j)$ is the "bilinear form"

$$J(f_i, f_j) = \iiint (f_i' f_j' - f_i f_j) g_{ij} b\, db\, d\epsilon\, d\mathbf{v}_j \tag{28-9}$$

representing the collision integrals.

The series solutions to the Boltzmann equations are obtained by introducing a perturbation parameter ζ into the Boltzmann equation in such a manner that the frequency of collisions can be varied in an arbitrary manner without affecting the relative number of collisions of a particular kind. Thus we consider a hypothetical problem in which the Boltzmann equation is

$$\frac{\partial f_i}{\partial t} + (\mathbf{v}_i \cdot \nabla f_i) + \frac{1}{m_i}(\mathbf{X}_i \cdot \nabla_{v_i} f_i) = \frac{1}{\zeta}\sum_j J(f_i, f_j) \tag{28-10}$$

and $1/\zeta$ measures the frequency of collisions. If ζ were small, collisions would be very frequent and the gas would behave like a continuum in which local equilibrium is everywhere maintained. The distribution function is expanded in a series in ζ,

$$f_i = f_i^{(0)} + \zeta f_i^{(1)} + \zeta^2 f_i^{(2)} + \cdots \tag{28-11}$$

If this series is introduced into the modified Boltzmann equation (28-10) and the coefficients of equal powers of ζ equated, one obtains the following set of equations for the functions, $f_i^{(0)}, f_i^{(1)}, f_i^{(2)}, \ldots$:

$$\sum_j J(f_i^{(0)}, f_j^{(0)}) = 0$$

$$\frac{\partial f_i^{(0)}}{\partial t} + (\mathbf{v}_i \cdot \nabla f_i^{(0)}) + \frac{1}{m_i}(\mathbf{X}_i \cdot \nabla_{v_i} f_i^{(0)}) = \sum_j [J(f_i^{(0)}, f_j^{(1)}) + J(f_i^{(1)}, f_j^{(0)})]$$

$$\frac{\partial f_i^{(1)}}{\partial t} + (\mathbf{v}_i \cdot \nabla f_i^{(1)}) + \frac{1}{m_i}(\mathbf{X}_i \cdot \nabla_{v_i} f_i^{(1)}) = \sum_j \begin{bmatrix} J(f_i^{(0)}, f_j^{(2)}) + J(f_i^{(1)}, f_j^{(1)}) \\ + J(f_i^{(2)}, f_j^{(0)}) \end{bmatrix}$$

$$\cdots \tag{28-12}$$

These equations for the $f_i^{(r)}$ serve to determine the distribution function uniquely in the manner described below.

The first expression of Eq. 28-12 is the set of coupled integral equations considered in the discussion of the equilibrium solution of the

Boltzmann equation and the H-theorem. From the arguments presented there, it is clear that the most general solution of these equations is

$$f_i^{(0)} = n_i \left(\frac{m_i}{2\pi kT}\right)^{\frac{3}{2}} e^{-m_i(\mathbf{v}_i-\mathbf{v}_0)^2/2kT} \tag{28-13}$$

The quantities

$$n_i = n_i(\mathbf{r}, t), \qquad \mathbf{v}_0 = \mathbf{v}_0(\mathbf{r}, t), \qquad T = T(\mathbf{r}, t) \tag{28-14}$$

are arbitrary functions of space and time insofar as this set of equations is concerned. In order for these functions to represent the local values of the physical quantities, i.e. number density, mass average velocity, and temperature, it is necessary that the solutions of the remaining equations be such that

$$\int f_i d\mathbf{v}_i = n_i \tag{28-15}$$

$$\sum_i m_i \int \mathbf{v}_i f_i d\mathbf{v}_i = \rho \mathbf{v}_0 \tag{28-16}$$

$$\tfrac{1}{2} \sum_i m_i \int (\mathbf{v}_i - \mathbf{v}_0)^2 f_i d\mathbf{v}_i = \tfrac{3}{2} nkT \tag{28-17}$$

It may be shown [*38*] that the remaining integral equations of Eq. 28-12, along with the auxiliary conditions,

$$\int f_i^{(r)} d\mathbf{v}_i = 0 \quad r = 1, 2, 3, \cdots \tag{28-18}$$

$$\sum_i m_i \int \mathbf{v}_i f_i^{(r)} d\mathbf{v}_i = 0 \quad r = 1, 2, 3, \cdots \tag{28-19}$$

$$\tfrac{1}{2} \sum_i m_i \int (\mathbf{v}_i - \mathbf{v}_0)^2 f_i^{(r)} d\mathbf{v}_i = 0 \quad r = 1, 2, 3, \cdots \tag{28-20}$$

specify uniquely a set of functions $f_i^{(r)}$. Since the set of distribution functions so defined satisfies the conditions of Eq. 28-15, 16, and 17, it is used as the solution to the Boltzmann equation.

The first order perturbation solution. The equation for the $f_i^{(1)}$ may be written in terms of a perturbation function ϕ_i, defined such that

$$f_i^{(1)}(\mathbf{r}, \mathbf{v}_i, t) = f_i^{(0)}(\mathbf{r}, \mathbf{v}_i, t)\phi_i(\mathbf{r}, \mathbf{v}_i, t) \tag{28-21}$$

In terms of this function the second of Eq. 28-12 becomes

$$\frac{\partial f_i^{(0)}}{\partial t} + (\mathbf{v}_i \cdot \nabla f_i^{(0)}) + \frac{1}{m_i}(\mathbf{X}_i \cdot \nabla_{v_i} f_i^{(0)})$$
$$= \sum_j \iiint f_i^{(0)} f_j^{(0)} (\phi_i' + \phi_j' - \phi_i - \phi_j) g_{ij} b\, db\, d\epsilon\, d\mathbf{v}_j \tag{28-22}$$

The preceding auxiliary conditions, in terms of ϕ_i are:

$$\int f_i^{(0)}\phi_i d\mathbf{v}_i = 0 \qquad (28\text{-}23)$$

$$\sum_i m_i \int \mathbf{v}_i f_i^{(0)}\phi_i d\mathbf{v}_i = 0 \qquad (28\text{-}24)$$

$$\tfrac{1}{2}\sum_i m_i \int (\mathbf{v}_i - \mathbf{v}_0)^2 f_i^{(0)}\phi_i d\mathbf{v}_i = 0 \qquad (28\text{-}25)$$

As stated above this set of equations is just sufficient to define the perturbation functions ϕ_i uniquely.

The differentiations of the function $f_i^{(0)}$ indicated in Eq. 28-22 may be carried out. The resulting expressions involve space and time derivatives of the functions, $n_i(\mathbf{r}, t)$, $\mathbf{v}_0(\mathbf{r}, t)$, and $T(\mathbf{r}, t)$. The time derivatives are eliminated by means of the equations of change (27-40, 43, 48). It is consistent with this approximation to replace f_i by $f_i^{(0)}$ in the integrals for the flux vectors which occur in the equations of change (that is, we use $\overline{\mathbf{V}}_i = 0$, $\mathsf{p} = p\mathsf{1}$, $\mathbf{q} = 0$). The resulting equation for the perturbation function ϕ_i is

$$f_i^{(0)}\left[\frac{n}{n_i}(\mathbf{V}_i \cdot \mathbf{d}_i) + (\mathsf{b}_i : \nabla \mathbf{v}_0) - \left(\frac{5}{2} - W_i^2\right)(\mathbf{V}_i \cdot \nabla \ln T)\right]$$
$$= \sum_j \iiint f_i^{(0)} f_j^{(0)}(\phi_i' + \phi_j' - \phi_i - \phi_j) g_{ij} b\, db\, d\epsilon\, d\mathbf{v}_j \qquad (28\text{-}26)$$

The quantities $\mathbf{d}_i$ and b_i are defined by

$$\mathbf{d}_i = \nabla\left(\frac{n_i}{n}\right) + \left(\frac{n_i}{n} - \frac{n_i m_i}{\rho}\right)\nabla \ln p - \left(\frac{n_i m_i}{p\rho}\right)\left[\frac{\rho}{m_i}\mathbf{X}_i - \sum_j n_j \mathbf{X}_j\right] \qquad (28\text{-}27)$$

$$\mathsf{b}_i = 2[\mathbf{W}_i\mathbf{W}_i - \tfrac{1}{3}W_i^2\mathsf{1}] \qquad (28\text{-}28)$$

The dimensionless velocity $\mathbf{W}_i$ is defined by Eq. 27-9.

The integral equations. The perturbation function ϕ_i depends upon space and time only through the quantities n_i, $\mathbf{v}_0$, and T and their space derivatives. It is clear from the form of the integral equation for ϕ_i that this quantity is linear in the derivatives and has the form

$$\phi_i = -(\mathbf{A}_i \cdot \nabla \ln T) - (\mathsf{B}_i : \nabla \mathbf{v}_0) + n \sum_j (\mathbf{C}_i^{(j)} \cdot \mathbf{d}_j) \qquad (28\text{-}29)$$

where the $\mathbf{A}_i$, B_i, and $\mathbf{C}_i^{(j)}$ are functions of the dimensionless velocity $\mathbf{W}_i$, the local composition, and the local temperature. If there are ν components to the chemical mixture, there are only $(\nu - 1)$ independent vectors $\mathbf{d}_i$, since according to the definition of the $\mathbf{d}_i$

$$\sum_i \mathbf{d}_i = 0 \tag{28-30}$$

This fact enables us to set one of the $\mathbf{C}_i^{(j)}$ equal to zero for each i. To retain the symmetry we let

$$\mathbf{C}_i^{(i)} = 0 \tag{28-31}$$

When the expression for ϕ_i in Eq. 28-29 is inserted into Eq. 28-26 and the coefficients of similar gradients are equated, there result separate integral equations for the functions $\mathbf{C}_i^{(j)}$, B_i, and $\mathbf{A}_i$,[45]

$$\frac{1}{n_i} f_i^{(0)}(\delta_{ih} - \delta_{ik})\mathbf{V}_i = \sum_j \cdot \iiint (\mathbf{C}_i^{(h)\prime} + \mathbf{C}_j^{(h)\prime} - \mathbf{C}_i^{(k)\prime}$$
$$- \mathbf{C}_j^{(k)\prime} - \mathbf{C}_i^{(h)} - \mathbf{C}_j^{(h)} + \mathbf{C}_i^{(k)} + \mathbf{C}_j^{(k)}) f_i^{(0)} f_j^{(0)} g_{ij} b\,db\,d\epsilon\,d\mathbf{v}_j \tag{28-32}$$

$$f_i^{(0)}\mathsf{b}_i = -\sum_j \iiint (\mathsf{B}_i' + \mathsf{B}_j' - \mathsf{B}_i - \mathsf{B}_j) f_i^{(0)} f_j^{(0)} g_{ij} b\,db\,d\epsilon\,d\mathbf{v}_j \tag{28-33}$$

$$f_i^{(0)}(\tfrac{5}{2} - W_i^2)\mathbf{V}_i = \sum_j \iiint (\mathbf{A}_i' + \mathbf{A}_j' - \mathbf{A}_i - \mathbf{A}_j) f_i^{(0)} f_j^{(0)} g_{ij} b\,db\,d\epsilon\,d\mathbf{v}_j \tag{28-34}$$

---

[45] Eq. 28-32 is obtained in the following way: We have by inserting from Eq. 28-29 into Eq. 28-26

$$\frac{1}{n_i} f_i^{(0)}(\mathbf{V}_i \cdot \mathbf{d}_i) = \sum_{j,h} \iiint [(\mathbf{C}_i^{(h)\prime} + \mathbf{C}_j^{(h)\prime} - \mathbf{C}_i^{(h)} - \mathbf{C}_j^{(h)}) \cdot \mathbf{d}_h] f_i^{(0)} f_j^{(0)} g_{ij} b\,db\,d\epsilon\,d\mathbf{v}_j$$

Then making use of the fact that

$$\sum_j \mathbf{d}_j = 0$$

and making algebraic rearrangements,

$$\frac{1}{n_i} f_i^{(0)} \left(\mathbf{V}_i \cdot \sum_h \delta_{ih} \mathbf{d}_h \right)$$
$$= \sum_{h \neq k} \mathbf{d}_h \sum_j \iiint \begin{bmatrix} \mathbf{C}_i^{(h)\prime} + \mathbf{C}_j^{(h)\prime} - \mathbf{C}_i^{(k)\prime} - \mathbf{C}_j^{(k)\prime} \\ -\mathbf{C}_i^{(h)} - \mathbf{C}_j^{(h)} + \mathbf{C}_i^{(k)} + \mathbf{C}_j^{(k)} \end{bmatrix} f_i^{(0)} f_j^{(0)} g_{ij} b\,db\,d\epsilon\,d\mathbf{v}_j$$

The left-hand side may now be rewritten as

$$\frac{1}{n_i} f_i^{(0)} \left(\mathbf{V}_i \cdot \sum_{h \neq k} (\delta_{ih} - \delta_{ik}) \mathbf{d}_h \right)$$

and coefficients of $\mathbf{d}_h$ may now be equated to give Eq. 28-32.

In these equations the only variables involved are $n_i(\mathbf{r}, t)$, $T(\mathbf{r}, t)$, and $\mathbf{V}_i(\mathbf{r}, t)$. (Actually we choose to work with $\mathbf{W}_i(\mathbf{r}, t)$ rather than $\mathbf{V}_i(\mathbf{r}, t)$.) The spatial coordinates (or derivatives) do not occur explicitly. Consequently, $\mathbf{A}_i$, $\mathbf{B}_i$, and $\mathbf{C}_i^{(j)}$ are functions of the space coordinates only through the variables mentioned above. It may be shown that any vector function of $\mathbf{W}_i$, the only vector variable, is the vector itself multiplied by some scalar function of the absolute value of the vector $\mathbf{W}_i$. Hence $\mathbf{A}_i$ and $\mathbf{C}_i^{(j)}$ are of the form

$$\mathbf{C}_i^{(j)} = \mathbf{W}_i C_i^{(j)}(W_i) \tag{28-35}$$

$$\mathbf{A}_i = \mathbf{W}_i A_i(W_i) \tag{28-36}$$

It may also be shown that the only tensor $\mathbf{B}_i$, which is consistent with the form of the integral equation for $\mathbf{B}_i$, Eq. 28-33, is of the form

$$\mathbf{B}_i = [\mathbf{W}_i\mathbf{W}_i - \tfrac{1}{3}W_i^2\mathbf{1}]B_i(W_i) \tag{28-37}$$

The integral equations for the $\mathbf{A}_i$, $\mathbf{B}_i$, and $\mathbf{C}_i^{(j)}$ (28-32, 33, and 34), are, of course, to be solved in conjunction with the auxiliary conditions imposed by Eq. 28-23, 24, and 25. In terms of these quantities the auxiliary relations become

$$\sum_i \sqrt{m_i} \int [(\mathbf{C}_i^{(j)} - \mathbf{C}_i^{(k)}) \cdot \mathbf{W}_i] f_i^{(0)} d\mathbf{v}_i = 0 \tag{28-38}$$

$$\sum_i \sqrt{m_i} \int (\mathbf{A}_i \cdot \mathbf{W}_i) f_i^{(0)} d\mathbf{v}_i = 0 \tag{28-39}$$

There is no auxiliary equation for the $\mathbf{B}_i$ analogous to these equations inasmuch as the functions of the form defined by Eq. 28-37 automatically satisfy the constraints given in Eq. 28-23, 24, and 25 for any arbitrary function $B_i(W_i)$.

Solutions to these integral equations have been obtained by two equivalent methods, that of Chapman and Cowling [*38*] and a variational method [*42*]. In both methods the scalar functions $C_i^{(j)}(W_i)$, $B_i(W_i)$, and $A_i(W_i)$ are expanded in a series of Sonine polynomials.[46] Chapman and Cowling used an *infinite series of these polynomials* with the result that the transport coefficients are expressed in terms of ratios of infinite determinants. To get numerical values it is necessary to consider only a few elements of these determinants since the convergence of the ratios of determinants is quite rapid as additional rows and columns are included. Here we discuss the problem from the standpoint of the variational method.

Because the integral equations given in Eq. 28-32, 33, and 34 are quite similar in form, it is possible to write one general equation for a

[46] See Eq. 28-57 for the definition of these polynomials.

tensor $\mathsf{T}_i^{(h,k)}$, which includes all three of these equations,[47]

$$\mathsf{R}_i^{(h,k)} = \sum_j \iiint [\mathsf{T}_i^{(h,k)\prime} + \mathsf{T}_j^{(h,k)\prime} - \mathsf{T}_i^{(h,k)} - \mathsf{T}_j^{(h,k)}] f_i^{(0)} f_j^{(0)} g_{ij} b\,db\,d\epsilon\,d\mathbf{v}_j \qquad (28\text{-}40)$$

The correspondence between the symbols $\mathsf{R}_i^{(h,k)}$ and $\mathsf{T}_i^{(h,k)}$, and their counterparts in Eq. 28-32, 33, and 34 is given in the following table:

Equation	$\mathsf{R}_i^{(h,k)}$	$\mathsf{T}_i^{(h,k)}$
(28-32)	$\frac{1}{n_i} f_i^{(0)}(\delta_{ih} - \delta_{ik})\mathbf{V}_i$	$\mathsf{C}_i^{(h)} - \mathsf{C}_i^{(k)}$
(28-33)	$-2f_i^{(0)}(\mathbf{W}_i\mathbf{W}_i - \frac{1}{3}W_i^2\mathbf{1})$	B_i
(28-34)	$f_i^{(0)}(\frac{5}{2} - W_i^2)\mathbf{V}_i$	$\mathbf{A}_i$

It should be kept in mind that the subscript i in the symbol $\mathsf{T}_i^{(h,k)}$ indicates that the tensor is a function of the velocity vector $\mathbf{W}_i$. The subscripts on the other symbols, R, A, B, C, have the same significance. We shall seek an approximate solution to the integral equation for $\mathsf{T}_i^{(h,k)}$, Eq. 28-40, by a variational method. This equation, together with the auxiliary equation

$$\sum_i \sqrt{m_i} \int (\mathsf{T}_i^{(h,k)} \cdot \mathbf{W}_i) f_i^{(0)} d\mathbf{v}_i = 0 \qquad (28\text{-}41)$$

(which is equivalent to Eq. 28-38 and 39) serves to specify $\mathsf{T}_i^{(h,k)}$ uniquely.

Several important integral theorems. In the following development, several abbreviations in the notation will be useful. Let G_{ij} and H_{ij} be any two tensors, in general functions of both $\mathbf{W}_i$ and $\mathbf{W}_j$. Then let us define $[\mathsf{G}_{ij}; \mathsf{H}_{ij}]_{ij}$ by[48]

$$[\mathsf{G}_{ij}; \mathsf{H}_{ij}]_{ij} = -\frac{1}{n_i n_j} \iiiint (\mathsf{G}_{ij} : (\mathsf{H}'_{ij} - \mathsf{H}_{ij})) f_i^{(0)} f_j^{(0)} g_{ij} b\,db\,d\epsilon\,d\mathbf{v}_i d\mathbf{v}_j \qquad (28\text{-}42)$$

From symmetry arguments similar to those introduced in Art. 27 it follows that

$$[\mathsf{G}_{ij}; \mathsf{H}_{ij}]_{ij} = \frac{1}{2n_i n_j} \iiiint ((\mathsf{G}'_{ij} - \mathsf{G}_{ij}) : (\mathsf{H}'_{ij} - \mathsf{H}_{ij})) f_i^{(0)} f_j^{(0)} g_{ij} b\,db\,d\epsilon\,d\mathbf{v}_i d\mathbf{v}_j \qquad (28\text{-}43)$$

Hence the bracket is symmetrical with respect to the interchange of G_{ij} and H_{ij} and also of the i and j subscripts on the bracket

$$[\mathsf{G}_{ij}; \mathsf{H}_{ij}]_{ij} = [\mathsf{H}_{ij}; \mathsf{G}_{ij}]_{ij} = [\mathsf{G}_{ij}; \mathsf{H}_{ij}]_{ji} = [\mathsf{H}_{ij}; \mathsf{G}_{ij}]_{ji} \qquad (28\text{-}44)$$

[47] Two of the equations involve the vector quantities (tensors of order 1) $\mathbf{A}_i$ and $\mathsf{C}_i^{(h)} - \mathsf{C}_i^{(k)}$.

[48] The subscript ij on the bracket indicates an integration over the variables $\mathbf{v}_i$ and $\mathbf{v}_j$.

This "square bracket" is a linear operator. If G_{ij} and H_{ij} have the form

$$\mathsf{G}_{ij} = \mathsf{K}_i + \mathsf{L}_j, \qquad \mathsf{H}_{ij} = \mathsf{M}_i + \mathsf{N}_j \tag{28-45}$$

where K_i and M_i depend only on $\mathbf{W}_i$, and L_j and N_j depend only on $\mathbf{W}_j$, then it is clear that

$$\begin{aligned}[\mathsf{K}_i + \mathsf{L}_j; \mathsf{M}_i + \mathsf{N}_j]_{ij} &= [\mathsf{K}_i; \mathsf{M}_i + \mathsf{N}_j]_{ij} + [\mathsf{L}_j; \mathsf{M}_i + \mathsf{N}_j]_{ij} \\ &= [\mathsf{K}_i; \mathsf{M}_i]_{ij} + [\mathsf{K}_i; \mathsf{N}_j]_{ij} \\ &\qquad + [\mathsf{L}_j; \mathsf{M}_i]_{ij} + [\mathsf{L}_j; \mathsf{N}_j]_{ij} \end{aligned} \tag{28-46}$$

It should be noted that the subscripts on the symbols within the bracket indicate the functional dependence of the tensors on the velocities $\mathbf{W}_i$ and $\mathbf{W}_j$. The subscript ij on the bracket itself indicates that the integral is evaluated for collisions between molecules i and j.

Let us consider two sets of tensor functions K_i and L_i and define an additional quantity in terms of these sets by

$$\{\mathsf{K}; \mathsf{L}\} = \sum_{i,j} n_i n_j [\mathsf{K}_i + \mathsf{K}_j; \mathsf{L}_i + \mathsf{L}_j]_{ij} \tag{28-47}$$

These "curly brackets" satisfy the following relations:

$$\{\mathsf{K}; \mathsf{L}\} = \{\mathsf{L}; \mathsf{K}\} \tag{28-48}$$

$$\{\mathsf{K}; \mathsf{L} + \mathsf{M}\} = \{\mathsf{K}; \mathsf{L}\} + \{\mathsf{K}; \mathsf{M}\} \tag{28-49}$$

Inasmuch as $\{\mathsf{K}; \mathsf{K}\}$ represents a sum of integrals, all of which have non-negative integrands, it follows that

$$\{\mathsf{K}; \mathsf{K}\} \geqq 0 \tag{28-50}$$

It can easily be shown that the curly bracket $\{\mathsf{K}; \mathsf{K}\}$ vanishes if and only if K_i is a linear combination of the constants of motion. The only linear combination of the constants of motion which satisfies the auxiliary condition, Eq. 28-41, is identically zero. Hence, if we restrict the consideration to sets of tensor functions K_i which satisfy the auxiliary condition, it follows that the curly bracket $\{\mathsf{K}; \mathsf{K}\}$ vanishes if and only if each of the K_i are identically zero.

A variational principle. A variational principle may be employed now to obtain approximate solutions to the integral equations, Eq. 28-40. Let us use as trial functions a set of functions $\mathsf{t}_i^{(h,k)}$, which satisfy the equations

$$\int (\mathsf{t}_i^{(h,k)} : \mathsf{R}_i^{(h,k)}) d\mathbf{v}_i = -\sum_j n_i n_j [\mathsf{t}_i^{(h,k)}; \mathsf{t}_i^{(h,k)} + \mathsf{t}_j^{(h,k)}]_{ij} \tag{28-51}$$

and which contain as many arbitrary parameters as is convenient.

If $\mathsf{t}_i^{(h,k)}$ is "double-dotted" into the integral equation for $\mathsf{T}_i^{(h,k)}$ (Eq.

28-40) and an integration carried out over $\mathbf{v}_i$, one obtains

$$\int (\mathbf{t}_i^{(h,k)}:\mathbf{R}_i^{(h,k)})d\mathbf{v}_i = -\sum_j n_i n_j[\mathbf{t}_i^{(h,k)}; \mathbf{T}_i^{(h,k)} + \mathbf{T}_j^{(h,k)}]_{ij} \qquad (28\text{-}52)$$

where the $\mathbf{t}_i^{(h,k)}$ are the trial functions and the $\mathbf{T}_i^{(h,k)}$ are the exact solutions to the integral equations. Equating the right sides of the last two equations and summing over i, one obtains (after making use of the symmetry relations, Eq. 28-44 and the definition of the curly bracket, Eq. 28-47),

$$\{\mathbf{t}^{(h,k)}; \mathbf{T}^{(h,k)}\} = \{\mathbf{t}^{(h,k)}; \mathbf{t}^{(h,k)}\} \qquad (28\text{-}53)$$

Since the curly bracket of identical sets of functions is non-negative (Eq. 28-50),

$$\{\mathbf{T}^{(h,k)} - \mathbf{t}^{(h,k)}; \mathbf{T}^{(h,k)} - \mathbf{t}^{(h,k)}\} \geqq 0 \qquad (28\text{-}54)$$

Then making use of the linear operator property of the curly brackets, Eq. 28-49, and the relation between the trial functions and the exact solutions, Eq. 28-53, we find that,

$$\{\mathbf{t}^{(h,k)}; \mathbf{t}^{(h,k)}\} \leqq \{\mathbf{T}^{(h,k)}; \mathbf{T}^{(h,k)}\} \qquad (28\text{-}55)$$

This is the statement of the variational method of obtaining approximations to the solutions $\mathbf{T}_i^{(h,k)}$. Specifically, the method of solution is as follows: One begins by choosing a set of trial functions, $\mathbf{t}_i^{(h,k)}$, which contain a number of arbitrary parameters. Then if only those trial functions are considered which satisfy the auxiliary condition, Eq. 28-41, the equality sign in Eq. 28-55 applies only when $\mathbf{T}_i^{(h,k)}$ and $\mathbf{t}_i^{(h,k)}$ are identical. Thus the best approximation to the true solution of the integral equation is obtained by maximizing the curly bracket on the left of Eq. 28-55 with respect to all of the available parameters in the set of trial functions. That is, for the best approximation

$$\delta\{\mathbf{t}^{(h,k)}; \mathbf{t}^{(h,k)}\} = -2\delta\sum_i \int (\mathbf{t}_i^{(h,k)}:\mathbf{R}_i^{(h,k)})d\mathbf{v}_i = 0 \qquad (28\text{-}56)$$

This, along with Eq. 28-51, which restricts the choice of the trial functions, forms the basis of the variational method of solution of the integral equations, Eq. 28-40.

The application of the variational principle. Let us now consider the implications of the variational principle discussed above when the trial functions are taken to be finite series of polynomials in the square of the velocity, W_i^2. It is convenient to make use of the Sonine polynomials, $S_n^{(m)}$, defined by[49]

[49] Except for the normalization these polynomials are the same as the associated Laguerre polynomials. The first two Sonine polynomials are

$$S_m^{(0)}(x) = 1$$

$$S_m^{(1)}(x) = m + 1 - x$$

$$S_n^{(m)}(x) = \sum_j \frac{(-1)^j(m+n)!}{(n+j)!(m-j)!j!} x^j \tag{28-57}$$

These polynomials satisfy the orthogonality condition,

$$\int_0^\infty x^n e^{-x} S_n^{(m)}(x) S_n^{(m')}(x)\,dx = \frac{(n+m)!}{m!} \delta_{mm'} \tag{28-58}$$

and are convenient, for as two special cases of this orthogonality relation we have

$$\int f_i^{(0)} S_{\frac{3}{2}}^{(m)}(W_i^2) V_i^2 d\mathbf{V}_i = \frac{3n_i kT}{m_i} \delta_{m0} \tag{28-59}$$

$$\int f_i^{(0)} S_{\frac{3}{2}}^{(m)}(W_i^2) V_i^4 d\mathbf{V}_i = 15 n_i \left(\frac{kT}{m_i}\right)^2 \delta_{m0} \tag{28-60}$$

As the trial functions, $\mathbf{t}_i^{(h,k)}$, we now take a *finite* linear combination of the Sonine polynomials, $S_n^{(m)}(W_i^2)$,

$$\mathbf{t}_i^{(h,k)}(\mathbf{W}_i) = \mathbb{W}_i \sum_{m=0}^{\xi-1} t_{im}^{(h,k)}(\xi) S_n^{(m)}(W_i^2) \tag{28-61}$$

in which the values of the index, n, and the meaning of the tensor $\mathbb{W}_i$ are as follows:

When $\mathbf{T}_i^{(h,k)}$ is	The value of the index, n is	The quantity $\mathbb{W}_i$ is	The Sonine expansion coefficients $t_{im}^{(h,k)}$ will be designated by
$\mathbf{A}_i$	$\frac{3}{2}$	$\mathbf{W}_i$	$a_{im}(\xi)$
$\mathbf{B}_i$	$\frac{5}{2}$	$\mathbf{W}_i\mathbf{W}_i - \frac{1}{3}W_i^2\mathbf{1}$	$b_{im}(\xi)$
$\mathbf{C}_i^{(h)} - \mathbf{C}_i^{(k)}$	$\frac{3}{2}$	$\mathbf{W}_i$	$c_{im}^{(h,k)}(\xi)$

For reasons which will become apparent later we have indicated the dependence of the expansion coefficients on the number of terms, ξ, used in the finite series trial function. These coefficients are *not* coefficients in an *infinite series* expansion and therefore do depend upon the number of terms used.

Let us define

$$R_{im}^{(h,k)} = \int (\mathbf{R}_i^{(h,k)} : \mathbb{W}_i) S_n^{(m)}(W_i^2)\,d\mathbf{V}_i \tag{28-62}$$

and

$$g^{(h,k)} = \sum_{i,m} t_{im}^{(h,k)} R_{im}^{(h,k)} \tag{28-63}$$

In terms of these quantities the constraints on the trial functions, Eq. 28-51, become

$$w_i^{(h,k)} = 0 \tag{28-64}$$

where

$$w_i^{(h,k)} = \sum_{m=0}^{\xi-1} t_{im}^{(h,k)} R_{im}^{(h,k)}$$

$$+ \sum_j \sum_{m=0}^{\xi-1} \sum_{m'=0}^{\xi-1} n_i n_j t_{im}^{(h,k)} \left[\begin{aligned} & t_{im'}^{(h,k)} [\mathbf{W}_i S_n^{(m)}(W_i^2); \mathbf{W}_i S_n^{(m')}(W_i^2)]_{ij} \\ & + t_{jm'}^{(h,k)} [\mathbf{W}_i S_n^{(m)}(W_i^2); \mathbf{W}_j S_n^{(m)'}(W_j^2)]_{ij} \end{aligned} \right] \quad (28\text{-}65)$$

and the statement of the variational criterion, Eq. 28-56, becomes

$$\delta g^{(h,k)} = 0 \quad (28\text{-}66)$$

The problem then is to find the extremum of $g^{(h,k)}$ subject to the constraints of Eq. 28-64. This extremum is determined by the method of the Lagrangian multipliers. Let $\lambda_i^{(h,k)}$ be the multipliers. Then Eq. 28-64 and the equations,

$$\left(\frac{\partial g^{(h,k)}}{\partial t_{im}^{(h,k)}} \right) + \sum_r \lambda_r^{(h,k)} \left(\frac{\partial w_r^{(h,k)}}{\partial t_{im}^{(h,k)}} \right) = 0 \quad (28\text{-}67)$$

are sufficient to determine the $\lambda_i^{(h,k)}$ and the expansion coefficients $t_{im}^{(h,k)}(\xi)$. Performing the indicated differentiations, we get

$$[1 + \lambda_i^{(h,k)}] R_{im}^{(h,k)}$$

$$+ \sum_j \sum_{m'=0}^{\xi-1} n_i n_j \left[\begin{aligned} & 2\lambda_i^{(h,k)} t_{im'}^{(h,k)} [\mathbf{W}_i S_n^m(W_i^2); \mathbf{W}_i S_n^{(m')}(W_i^2)]_{ij} \\ & + (\lambda_i^{(h,k)} + \lambda_j^{(h,k)}) t_{jm'}^{(h,k)} [\mathbf{W}_i S_n^{(m)}(W_i^2); \mathbf{W}_j S_n^{(m')}(W_j^2)]_{ij} \end{aligned} \right] = 0$$

$$i = 1, 2, \cdots, \nu \quad (28\text{-}68)$$

$$m = 0, 1, \cdots, \xi - 1$$

The only solution to this set of equations together with Eq. 28-64 is

$$\lambda_i^{(h,k)} = 1 \quad i = 1, 2, 3, \cdots, \nu \quad (28\text{-}69)$$

with the constants $t_{im}^{(h,k)}(\xi)$ determined by Eq. 28-68, with $\lambda_i^{(h,k)} = 1$. The equations may be rewritten in the form,

$$\sum_j \sum_{m'=0}^{\xi-1} Q_{ij}^{mm'} t_{jm'}^{(h,k)}(\xi) = -R_{im}^{(h,k)} \quad (28\text{-}70)$$

where

$$Q_{ij}^{mm'} = \sum_l n_i n_l \left[\begin{aligned} & \delta_{ij} [\mathbf{W}_i S_n^{(m)}(W_i^2); \mathbf{W}_i S_n^{(m')}(W_i^2)]_{il} \\ & + \delta_{jl} [\mathbf{W}_i S_n^{(m)}(W_i^2); \mathbf{W}_l S_n^{(m')}(W_l^2)]_{il} \end{aligned} \right] \quad (28\text{-}71)$$

For the case where $\mathbf{T}_i^{(h,k)}$ is $\mathbf{B}_i$ (and $t_{im}^{(h,k)}$ is $b_{im}^{(h,k)}$) Eq. 28-70 are all linearly independent, and hence all the $b_{im}^{(h,k)}$ can be determined from these equations. However, when $\mathbf{T}_i^{(h,k)}$ is either $\mathbf{A}_i$ or $(\mathbf{C}_i^{(h)} - \mathbf{C}_i^{(k)})$ then it may be shown that the set of equations for $m = 0$ includes one redundant equation. For these cases we make use of the auxiliary relation (Eq. 28-41) which in terms of the $t_{im}^{(h,k)}$ is

$$\sum_i \sum_{m=0}^{\xi-1} \sqrt{m_i}\, t_{im}^{(h,k)}(\xi) \int W_i^2 S_{\frac{3}{2}}^{(m)}(W_i^2) f_i^{(0)} d\mathbf{V}_i = 0 \tag{28-72}$$

Because of the orthogonality relation (Eq. 28-59) the terms in this sum are zero for m other than zero. Thus the auxiliary condition becomes

$$\sum_i \sqrt{m_i}\, t_{i0}^{(h,k)}(2n_i/\sqrt{\pi})\Gamma(\tfrac{5}{2}) = 0 \tag{28-73}$$

Consequently, when $\mathbf{T}_i^{(h,k)}$ is either $\mathbf{A}_i$ or $(\mathbf{C}_i^{(h)} - \mathbf{C}_i^{(k)})$, the trial function must be chosen so that

$$\sum_i n_i \sqrt{m_i}\, t_{i0}^{(i,k)} = 0 \tag{28-74}$$

This supplies the additional information needed for the specification of all the coefficients $t_{im}^{(h,k)}(\xi)$. Eq. 28-70 may be modified to include this statement. The result is

$$\sum_j \sum_{m'=0}^{\xi-1} \bar{Q}_{ij}^{mm'} t_{jm'}^{(h,k)}(\xi) = -R_{im}^{(h,k)} \tag{28-75}$$

where

$$\bar{Q}_{ij}^{mm'} = \begin{cases} Q_{ij}^{mm'}, & \text{when } t_{jm'}^{(h,k)} = b_{jm'} \\ Q_{ij}^{mm'} - \dfrac{n_j\sqrt{m_j}}{n_i\sqrt{m_i}} Q_{ii}^{mm'}\delta_{m0}\delta_{m'0}, & \text{when } t_{jm'}^{(h,k)} = a_{jm'} \text{ or } c_{jm'}^{(h,k)} \end{cases} \tag{28-76}$$

From Eq. 28-75 one can obtain the $t_{im}^{(h,k)}(\xi)$, which give the functions $\mathbf{A}_i$, $\mathbf{B}_i$, and $\mathbf{C}_i^{(j)}$. These in turn give the function ϕ_i and the distribution function f_i correct to the first order. The knowledge of the nonequilibrium distribution function f_i then provides the information needed for the evaluation of the transport coefficients. Fortunately it turns out that only a few terms in the Sonine expansion are needed. For viscosity and diffusion, one term gives a good approximation—the result of using two terms differs by only a few per cent. For diffusion, the use of one term alone does not describe the dependence of the diffusion coefficient on concentration; the slight concentration-dependence is brought out when the first two terms in the Sonine expansion are used. The transfer of mass due to a temperature gradient (i.e. thermal diffusion) does not appear at all in the expression for the diffusion velocity if only one term in the Sonine polynomial expansion is used; thus the fact that thermal diffusion is a small effect compared with mass transfer due to a concentration gradient manifests itself in the mathematical formulation of the physical processes.

B,29. The Formulation of the Transport Coefficients. The integral expressions for the flux vectors, which describe the flux of mass, momentum, and energy are derived in Art. 27. The evaluation of these integrals requires a knowledge of the distribution function f_i. An approximation to the distribution function in the form $f_i = f_i^{(0)}(1 + \phi_i)$ is derived in the previous article. If the expression for ϕ_i (given by Eq. 28-29) is used in the integrals for the flux vectors, one obtains expressions for the diffusion velocity, the pressure tensor, and the heat flux vector in terms of integrals of the functions $A_i(W_i)$, $B_i(W_i)$, and $C_i^{(j)}(W_i)$. It is these relations for the flux vectors which we shall now derive and discuss.

Diffusion coefficients and thermal diffusion coefficients in terms of Sonine expansion coefficients. The integral for the diffusion velocity (27-17) rewritten in terms of ϕ_i is

$$\bar{\mathbf{V}}_i = \frac{1}{n_i}\int \mathbf{V}_i f_i d\mathbf{V}_i = \frac{1}{n_i}\int \mathbf{V}_i \phi_i f_i^{(0)} d\mathbf{V}_i \tag{29-1}$$

Using Eq. 28-29 and noting that the term involving $\mathbf{B}_i$ vanishes on integration, we may write $\bar{\mathbf{V}}_i$ as

$$\bar{\mathbf{V}}_i = \frac{1}{n_i}\int \left\{ n \sum_j (\mathbf{C}_i^{(j)} \cdot \mathbf{d}_j) - (\mathbf{A}_i \cdot \nabla \ln T) \right\} \mathbf{V}_i f_i^{(0)} d\mathbf{V}_i \tag{29-2}$$

Then making use of the form of the functions $\mathbf{A}_i$ and $\mathbf{C}_i^{(j)}$ indicated in Eq. 28-35 and 28-36, we obtain[50]

$$\bar{\mathbf{V}}_i = \left(\frac{n^2}{n_i\rho}\right) \sum_j m_j D_{ij} \mathbf{d}_j - \frac{1}{n_i m_i} D_i^T \nabla \ln T \tag{29-3}$$

In this expression $\mathbf{d}_j$ is defined in Eq. 28-27 and

$$D_{ij} = \frac{\rho}{3nm_j}\sqrt{\frac{2kT}{m_i}} \int C_i^{(j)}(W_i) W_i^2 f_i^{(0)} d\mathbf{V}_i \tag{29-4}$$

$$D_i^T = \tfrac{1}{3} m_i \sqrt{\frac{2kT}{m_i}} \int A_i(W_i) W_i^2 f_i^{(0)} d\mathbf{V}_i \tag{29-5}$$

are the "multicomponent diffusion coefficients" and the "multicomponent thermal diffusion coefficients" respectively.[51]

[50] Here we use a theorem which may be proved by symmetry arguments: If $F(r)$ is any function of the absolute value of $\mathbf{r}$, then

$$\int F(r)\mathbf{r}\mathbf{r}d\mathbf{r} = \tfrac{1}{3}\mathbf{1} \int F(r) r^2 d\mathbf{r}$$

[51] There is a considerable variation among authors in the nomenclature and definition of these quantities. The D_{ij} defined here are such that for a two-component mixture the D_{ij} reduce to the usual "binary diffusion coefficient" $\mathfrak{D}_{ij}$. The quantities D_i^T have not been previously defined. In the case of two components, the D_i^T defined here do *not* reduce to the "binary thermal diffusion coefficients" as defined in Chapman and Cowling.

Thus we see that the diffusion velocity $\bar{\mathbf{V}}_i$ contains terms proportional to the concentration gradient, the pressure gradient, the difference in the external forces acting on the various species of molecules, and the gradient in the temperature. Before the work of Chapman and Enskog, discussed in the previous article, the latter had been unknown theoretically and unobserved experimentally. Subsequent experiments of Chapman and coworkers revealed that the theoretical prediction of the phenomenon of thermal diffusion was quite correct. This is one of a number of historically interesting instances of the prediction of experimentally observable phenomena by rigorous theoretical analysis. Experiments for the measurement of D_{ij} are usually arranged so that the contributions to the diffusion velocity resulting from pressure gradients, external forces, and thermal gradients are negligible.

By means of the Sonine polynomial expansions (Eq. 28-61) the expressions for the diffusion coefficients become

$$D_{ij}(\xi) = \left(\frac{\rho}{3nm_j}\right)\sqrt{\frac{m_i}{2kT}}\sum_{m=0}^{\xi-1} c_{im}^{(j,i)}(\xi)\int V_i^2 S_{\frac{3}{2}}^{(m)}(W_i^2) f_i^{(0)} d\mathbf{V}_i \quad (29\text{-}6)$$

$$D_i^T(\xi) = \frac{m_i}{3}\sqrt{\frac{m_i}{2kT}}\sum_{m=0}^{\xi-1} a_{im}(\xi)\int V_i^2 S_{\frac{3}{2}}^{(m)}(W_i^2) f_i^{(0)} d\mathbf{V}_i \quad (29\text{-}7)$$

the argument ξ of $D_{ij}(\xi)$ and $D_i^T(\xi)$ being the number of terms used in the Sonine expansions. Subsequent use of Eq. 28-59 to evaluate the integrals then gives

$$D_{ij}(\xi) = \frac{\rho n_i}{2nm_j}\sqrt{\frac{2kT}{m_i}}\, c_{i0}^{(j,i)}(\xi) \quad (29\text{-}8)$$

$$D_i^T(\xi) = \frac{n_i m_i}{2}\sqrt{\frac{2kT}{m_i}}\, a_{i0}(\xi) \quad (29\text{-}9)$$

We have thus expressed D_{ij} and D_i^T in terms of the *zeroeth* Sonine expansion coefficient alone. No matter how many terms are used in the expansion (i.e. regardless of the value of ξ), it is only the zeroeth coefficient which remains after the integrations have been performed. However, the values of the coefficients $c_{i0}^{(j,i)}(\xi)$ and $a_{i0}(\xi)$ which are determined by Eq. 28-75 depend upon the number of terms considered in the polynomial.

In the case of D_{ij} letting $\xi = 1$ gives quite good results; if $\xi = 2$, one obtains a small correction term. Except in very unusual cases, it is unnecessary to use any approximations beyond $\xi = 2$. However, when $\xi = 1$ the coefficients D_i^T vanish identically. This arises because the function, $\mathbf{R}_i^{(h,k)} = f_i^{(0)}[\frac{5}{2} - W_i^2]\mathbf{V}_i$, when used in Eq. 28-62 with $n = \frac{3}{2}$, causes this integral to vanish since $(\frac{5}{2} - W_i^2)$ is the same as $S_{\frac{3}{2}}^{(1)}(W_i^2)$ and is therefore orthogonal to $S_{\frac{3}{2}}^{(0)}(W_i^2)$. Hence in order to get the coefficient of thermal diffusion, it is necessary to take at least two terms in the Sonine expansion (i.e. $\xi = 2$). For realistic potential functions, this approximation is in error by about 5 to 10 per cent.

The coefficient of viscosity in terms of the Sonine expansion coefficients. The integral for the pressure tensor (Eq. 27-21) in terms of the perturbation function ϕ_i is

$$\mathbf{p} = \sum_j m_j \int \mathbf{V}_j\mathbf{V}_j f_j d\mathbf{V}_j \tag{29-10}$$

$$= \sum_j m_j \left\{ \int \mathbf{V}_j\mathbf{V}_j f_j^{(0)} d\mathbf{V}_j + \int \mathbf{V}_j\mathbf{V}_j f_j^{(0)} \phi_j d\mathbf{V}_j \right\} \tag{29-11}$$

$$= p\mathbf{1} + \sum_j m_j \int \mathbf{V}_j\mathbf{V}_j f_j^{(0)} \phi_j d\mathbf{V}_j \tag{29-12}$$

where

$$p = nkT \tag{29-13}$$

is the equilibrium hydrostatic pressure at the local temperature and density. Then using the form of the perturbation function ϕ_i as given by Eq. 28-29 we find that

$$\mathbf{p} = p\mathbf{1} - \sum_j m_j \int \mathbf{V}_j\mathbf{V}_j (\mathbf{B}_j : \nabla \mathbf{v}_0) f_j^{(0)} d\mathbf{V}_j \tag{29-14}$$

The terms in $\mathbf{A}_i$ and $\mathbf{C}_i^{(j)}$ can be shown to be zero on the basis of symmetry arguments. Then using the form of the tensor $\mathbf{B}_i$ as given in Eq. 28-37 and further symmetry arguments we obtain

$$\mathbf{p} = p\mathbf{1} - \left[\frac{2}{15} \sum_j \frac{m_j^2}{2kT} \int B_j(W_j) V_j^4 f_j^{(0)} d\mathbf{V}_j \right] \mathbf{S} \tag{29-15}$$

where $\mathbf{S}$ is the "rate of shear tensor," defined by

$$S_{\alpha\beta} = \frac{1}{2} \left[\frac{\partial v_{0\beta}}{\partial x_\alpha} + \frac{\partial v_{0\alpha}}{\partial x_\beta} \right] - \frac{1}{3} \delta_{\alpha\beta} (\nabla \cdot \mathbf{v}_0) \tag{29-16}$$

The coefficient of viscosity μ is defined by the relation

$$\mathbf{p} = p\mathbf{1} - 2\mu\mathbf{S} \tag{29-17}$$

From Eq. 29-15 it follows that the coefficient of viscosity is

$$\mu = \frac{1}{15} \sum_j \frac{m_j^2}{2kT} \int B_j(W_j) V_j^4 f_j^{(0)} d\mathbf{V}_j \tag{29-18}$$

Then making use of the Sonine polynomial expansion (Eq. 28-61) we obtain

$$\mu(\xi) = \frac{1}{15} \sum_j \frac{m_j^2}{2kT} \sum_{m=0}^{\xi-1} b_{jm}(\xi) \int S_{\frac{5}{2}}^{(m)}(W_j^2) V_j^4 f_j^{(0)} d\mathbf{V}_j \qquad (29\text{-}19)$$

Again the argument of $\mu(\xi)$ indicates the order of the approximation. The orthogonality relation, Eq. 28-60, can now be employed to evaluate the integral and one obtains

$$\mu(\xi) = \tfrac{1}{2} kT \sum_j n_j b_{j0}(\xi) \qquad (29\text{-}20)$$

as the coefficient of viscosity of a mixture in terms of the Sonine expansion coefficients $b_{j0}(\xi)$. As in the case of ordinary diffusion, the first approximation is the predominant contribution. Actual calculations show that there is very little change in μ when additional terms in the expansion are used.

Thermal conductivity. The integral for the energy or heat flux vector $\mathbf{q}$ (Eq. 27-24) is given in terms of the perturbation function ϕ_i by

$$\mathbf{q} = \tfrac{1}{2} \sum_j m_j \int V_j^2 \mathbf{V}_j f_j^{(0)} \phi_j d\mathbf{V}_j \qquad (29\text{-}21)$$

When the expression for ϕ_i (Eq. 28-29) is used in this equation, the term containing $\mathbf{B}_j$ does not contribute and we have

$$\mathbf{q} = \tfrac{1}{2} \sum_j m_j \int \left\{ n \sum_k (\mathbf{C}_j^{(k)} \cdot \mathbf{d}_k) - (\mathbf{A}_j \cdot \nabla \ln T) \right\} V_j^2 \mathbf{V}_j f_j^{(0)} d\mathbf{V}_j \qquad (29\text{-}22)$$

The energy flux may be separated into two parts—that due to the flux of molecules relative to the mass velocity and that due to other causes. Using the forms of the functions $\mathbf{A}_i$ and $\mathbf{C}_i^{(j)}$ given by Eq. 28-35 and 28-36, we obtain

$$\begin{aligned} \mathbf{q} = {} & \tfrac{5}{2} kT \sum_j n_j \overline{\mathbf{V}}_j \\ & - kT \sum_j \int \left\{ n \sum_k C_j^{(k)}(W_j)(\mathbf{W}_j \cdot \mathbf{d}_j) - A_j(W_j)(\mathbf{W}_j \cdot \nabla \ln T) \right\} \\ & \qquad \times (\tfrac{5}{2} - W_j^2) \mathbf{V}_j f_j^{(0)} d\mathbf{V}_j \end{aligned} \qquad (29\text{-}23)$$

Then according to the symmetry arguments discussed above it follows that

$$\begin{aligned} \mathbf{q} = {} & \tfrac{5}{2} kT \sum_j n_j \overline{\mathbf{V}}_j - \lambda' \nabla T \\ & - \frac{\sqrt{2}}{3} n (kT)^{\frac{3}{2}} \sum_k \mathbf{d}_k \sum_j \frac{1}{\sqrt{m_j}} \int C_j^{(k)}(W_j) \left(\frac{5}{2} - W_j^2 \right) W_j^2 f_j^{(0)} d\mathbf{V}_j \end{aligned} \qquad (29\text{-}24)$$

where

$$\lambda' = -\frac{\sqrt{2}}{3} k \sqrt{kT} \sum_j \frac{1}{\sqrt{m_j}} \int A_j(W_j) \left(\frac{5}{2} - W_j^2\right) W_j^2 f_j^{(0)} d\mathbf{V}_j \quad (29\text{-}25)$$

The first term represents the flux of energy incident to mass transport; the second, that due to a temperature gradient; and the third, an additional effect due directly to the concentration gradients. The last term is analogous to the effect of temperature gradients on diffusion, i.e. thermal diffusion.

The expression for $\mathbf{q}$ can be rearranged as follows: From the integral equation for the $\mathbf{A}_i$ (Eq. 28-34) we may obtain the relation

$$\sqrt{\frac{2kT}{m_j}} \int C_j^{(k)}(W_j) \left(\frac{5}{2} - W_j^2\right) W_j^2 f_j^{(0)} d\mathbf{V}_j$$
$$= \sum_i \iiiint ([\mathbf{A}_i' + \mathbf{A}_j' - \mathbf{A}_i - \mathbf{A}_j] \cdot \mathbf{C}_j^{(k)}) f_i^{(0)} f_j^{(0)} g_{ij} b\,db\,d\epsilon\,d\mathbf{V}_i d\mathbf{V}_j \quad (29\text{-}26)$$

which can be used to rewrite the expression for $\mathbf{q}$ (Eq. 29-24). Then "symmetrizing" the sum over "i" and "j" and making use of the fact that $\sum_k \mathbf{d}_k = 0$, one obtains

$$\mathbf{q} = \tfrac{5}{2} kT \sum_j n_j \overline{\mathbf{V}}_j - \lambda' \nabla T$$
$$- \tfrac{1}{8} nkT \sum_k \mathbf{d}_k \sum_{i,j} \iiiint ([\mathbf{A}_i' + \mathbf{A}_j' - \mathbf{A}_i - \mathbf{A}_j] \cdot [\mathbf{C}_i^{(k)} + \mathbf{C}_j^{(k)} - \mathbf{C}_i^{(h)} - \mathbf{C}_j^{(h)}])$$
$$f_i^{(0)} f_j^{(0)} g_{ij} b\,db\,d\epsilon\,d\mathbf{V}_i d\mathbf{V}_j \quad (29\text{-}27)$$

But from the integral equation for $\mathbf{C}_i^{(k)}$ (Eq. 28-32) it follows that

$$(\delta_{ih} - \delta_{ik}) \frac{1}{n_i} \sqrt{\frac{2kT}{m_i}} \int A_i(W_i) W_i^2 f_i^{(0)} d\mathbf{V}_i$$
$$= \sum_j \iiiint \left(\mathbf{A}_i \cdot \begin{bmatrix} \mathbf{C}_i^{(h)\prime} + \mathbf{C}_j^{(h)\prime} - \mathbf{C}_i^{(k)\prime} - \mathbf{C}_j^{(k)\prime} \\ - \mathbf{C}_i^{(h)} - \mathbf{C}_j^{(h)} + \mathbf{C}_i^{(k)} + \mathbf{C}_j^{(k)} \end{bmatrix}\right) f_i^{(0)} f_j^{(0)} g_{ij} b\,db\,d\epsilon\,d\mathbf{V}_i d\mathbf{V}_j$$
$$(29\text{-}28)$$

Summing Eq. 29-28 over i and symmetrizing results in

$$\sum_i (\delta_{ih} - \delta_{ik}) \frac{1}{n_i} \sqrt{\frac{2kT}{m_i}} \int A_i(W_i) W_i^2 f_i^{(0)} d\mathbf{V}_i$$
$$= -\tfrac{1}{2} \sum_{i,j} \iiiint \left(\begin{bmatrix} \mathbf{A}_i' + \mathbf{A}_j' \\ - \mathbf{A}_i - \mathbf{A}_j \end{bmatrix} \cdot \begin{bmatrix} \mathbf{C}_i^{(k)} + \mathbf{C}_j^{(k)} \\ - \mathbf{C}_i^{(h)} - \mathbf{C}_j^{(h)} \end{bmatrix}\right) f_i^{(0)} f_j^{(0)} g_{ij} b\,db\,d\epsilon\,d\mathbf{V}_i d\mathbf{V}_j$$
$$(29\text{-}29)$$

Then with the definition of D_j^T (Eq. 29-5) and the fact that $\sum_k \mathbf{d}_k = 0$, the energy flux becomes

$$\mathbf{q} = \frac{5}{2} kT \sum_j n_j \mathbf{\bar{V}}_j - \lambda' \nabla T - nkT \sum_j \frac{1}{n_j m_j} D_j^T \mathbf{d}_j \tag{29-30}$$

The coefficient λ' is not the coefficient of thermal conductivity as it is usually defined. It is conventional to eliminate the gradients $\mathbf{d}_j$ from the expression for $\mathbf{q}$ by means of the equation for the diffusion velocities (Eq. 29-3). The energy flux is then given in terms of the diffusion velocities and the temperature gradient. Because of the thermal diffusion term in the expression for the diffusion velocity, a small term adds to λ' to result in the quantity λ which is the usual coefficient of thermal conductivity. The final expression for λ (Eq. 29-65) is derived later in this article.

If the functions $A_i(W_i)$ are expressed as a series of Sonine polynomials as in Eq. 28-61 the expression for λ' (Eq. 29-25) becomes

$$\lambda'(\xi) =$$

$$-\frac{\sqrt{2}}{3} k \sqrt{kT} \sum_j \sum_{m=0}^{\xi-1} \frac{1}{\sqrt{m_j}} a_{jm}(\xi) \int S_{\frac{3}{2}}^{(m)}(W_j^2) \left(\frac{5}{2} - W_j^2\right) W_j^2 f_j^{(0)} d\mathbf{V}_j \tag{29-31}$$

Then since

$$S_{\frac{3}{2}}^{(1)}(W_j^2) = \tfrac{5}{2} - W_j^2 \tag{29-32}$$

one obtains from the condition of orthogonality of the polynomials, (Eq. 28-58),

$$\lambda'(\xi) = -\frac{5}{4} k \sum_j n_j \sqrt{\frac{2kT}{m_j}}\, a_{j1}(\xi) \tag{29-33}$$

The integrals, $\Omega^{(l)}(s)$. The transport coefficients have been expressed in terms of the Sonine polynomial expansion coefficients. It will be recalled that these expansion coefficients are obtained by the solution of sets of simultaneous equations, Eq. 28-75. It can be seen that the expansion coefficients $t_{jm}^{(h,k)}(\xi)$, are complicated combinations of the square bracket integrals, which were defined by Eq. 28-42. Chapman and Cowling [*38*] have shown that these integrals may be written as linear combinations of a set of integrals $\Omega^{(l)}(s)$. For collisions between molecules of type i and type j, these integrals (which are average collision cross sections) are defined by

$$\Omega_{ij}^{(l)}(s) = \sqrt{\pi} \left(\frac{m_{ij}}{2kT}\right)^{\frac{2s+3}{2}} \int_0^\infty \int_0^\infty e^{-\frac{m_{ij} g_{ij}^2}{2kT}} g_{ij}^{2s+3} (1 - \cos^l \chi) b\, db\, dg_{ij} \tag{29-34}$$

In these integrals m_{ij} is the reduced mass of colliding molecules, i and j,

g_{ij} is the initial relative speed of the colliding molecules, χ is the angle by which the molecules are deflected in the center of gravity coordinate system, and b is the impact parameter. A tabulation of the most frequently needed square brackets in terms of the $\Omega^{(l)}(s)$ has been given by Chapman and Cowling [*38*]. With these tables and Eq. 28-75 any of the Sonine expansion coefficients $t_{im}^{(h,k)}$ may be calculated.

The dynamics of the collisions enter into the description of the transport coefficients through the collision integrals defined by Eq. 29-34. For, in order to evaluate $\Omega^{(l)}(s)$, one must know χ as a function of the initial relative velocity g_{ij} and the impact parameter b. This relation depends on the form of the potential energy of interaction $\varphi(r)$. Thus given the interaction potential, one can calculate the $\Omega^{(l)}(s)$ integrals and hence the expansion coefficients $t_{im}^{(h,k)}(\xi)$.

The formulas for the transport coefficients in terms of the $\Omega^{(l)}(s)$ integrals are discussed in the next subarticle. The actual evaluation of the angles $\chi(g, b)$, the integrals $\Omega^{(l)}(s)$, and the transport coefficients for several intermolecular laws of force are discussed in Sec. D. The most satisfactory and usable calculations are those made on the basis of the Lennard-Jones potential, which describes reasonably well the interaction between spherical nonpolar molecules. Equations and tables are given in Sec. D which enable one to make practical applications of the theory. It is shown that the agreement between calculated and experimental results is quite satisfactory.

The transport coefficients in terms of the collision integrals. The method of obtaining formulas for the transport coefficients in terms of the intermolecular forces and the dynamics of binary collisions has been discussed in this and the preceding articles. The algebraic detail is rather lengthy and will be omitted, except to indicate briefly how the results obtained thus far may be used to obtain the lowest approximation to the various transport coefficients.

1. The Coefficient of Diffusion. The coefficient of diffusion in a multicomponent mixture can be obtained to a very good approximation by considering only one term in the Sonine polynomial expansion. The equations specifying the $c_{j0}^{(h,k)}(1)$ (Eq. 28-75) are then,

$$\sum_j \tilde{Q}_{ij}^{00} c_{j0}^{(h,k)}(1) = -R_{i0}^{(h,k)} \tag{29-35}$$

In terms of the $\Omega^{(l)}(s)$ these equations become

$$\sum_j c_{j0}^{(h,k)}(1) \sum_l \frac{n_l m_l}{(m_i + m_l)\sqrt{m_j}} [n_i m_i(\delta_{ij} - \delta_{jl}) - n_j m_j(1 - \delta_{il})]\Omega_{il}^{(1)}(1)$$
$$= -(\delta_{ih} - \delta_{ik}) \frac{3}{16} \sqrt{2kT} \tag{29-36}$$

For a binary mixture one immediately obtains from this equation,

$$c_{i0}^{(h,k)}(1) = (\delta_{jh} - \delta_{jk}) \frac{3}{16} \frac{(m_i + m_j)}{n_i\rho} \sqrt{\frac{2kT}{m_i}} \frac{1}{\Omega_{ij}^{(1)}(1)} \tag{29-37}$$

so that the first approximation to the coefficient of diffusion of a binary mixture[52] is (from Eq. 29-8)

$$\mathfrak{D}_{ij}(1) = \frac{3(m_i + m_j)}{16nm_im_j} \frac{kT}{\Omega_{ij}^{(1)}(1)} \tag{29-38}$$

From the definitions (or from the form of Eq. 29-37),

$$c_{j0}^{(h,k)} = c_{j0}^{(h,j)} - c_{j0}^{(k,j)} \tag{29-39}$$

Using this result, the expressions for the diffusion constants, and the above relation for the binary diffusion constants, we get for the equation for the general diffusion constants (Eq. 29-36)

$$\sum_j F_{ij}\{m_hD_{jh}(1) - m_kD_{jk}(1)\} = (\delta_{ih} - \delta_{ik}) \tag{29-40}$$

where

$$F_{ij} = \left\{\frac{n_i}{\rho\mathfrak{D}_{ij}(1)} + \sum_{l\neq i} \frac{n_lm_j}{\rho m_i\mathfrak{D}_{il}(1)}\right\} (1 - \delta_{ij}) \tag{29-41}$$

A formal solution of this set of equations can be obtained easily. Let us define F^{ij} as the cofactor of F_{ij} in the determinant $|F|$ of the F_{ij} so that

$$F^{ij} = (-1)^{i+j} \begin{vmatrix} F_{11} & \cdots & F_{1,j-1} & F_{1,j+1} & \cdots & F_{1\nu} \\ \cdot & & \cdot & \cdot & & \cdot \\ \cdot & & \cdot & \cdot & & \cdot \\ \cdot & & \cdot & \cdot & & \cdot \\ F_{i-1,1} & \cdots & F_{i-1,j-1} & F_{i-1,j+1} & \cdots & F_{i-1,\nu} \\ F_{i+1,1} & \cdots & F_{i+1,j-1} & F_{i+1,j+1} & \cdots & F_{i+1,\nu} \\ \cdot & & \cdot & \cdot & & \cdot \\ \cdot & & \cdot & \cdot & & \cdot \\ \cdot & & \cdot & \cdot & & \cdot \\ F_{\nu,1} & \cdots & F_{\nu,j-1} & F_{\nu,j+1} & \cdots & F_{\nu,\nu} \end{vmatrix} \tag{29-42}$$

Then solving Eq. 29-40,

$$m_hD_{ih}(1) - m_kD_{ik}(1) = \frac{F^{hi} - F^{ki}}{|F|} \tag{29-43}$$

Thus since $D_{ii} \equiv 0$, it follows that

$$D_{ij}(1) = \frac{F^{ji} - F^{ii}}{m_j|F|} \tag{29-44}$$

The set of Eq. 29-44 relates the generalized diffusion coefficients of a mixture to the binary diffusion coefficients of the various pairs. The form of the result, however, is usually difficult to handle in actual problems. For this reason it is often advantageous to make use of an alternate

[52] In a binary mixture, we denote D_{ij} by $\mathfrak{D}_{ij}$.

formulation. From Eq. 29-3, one finds that

$$\sum_j \frac{n_i n_j}{\mathfrak{D}_{ij}(1)} (\bar{\mathbf{V}}_i - \bar{\mathbf{V}}_j) = \frac{n^2}{\rho} \sum_{\substack{j,k \\ j \neq i}} \frac{m_k}{\mathfrak{D}_{ij}(1)} [n_j D_{ik} - n_i D_{jk}] \mathbf{d}_k$$

$$- (\nabla \ln T) \sum_j \frac{1}{\mathfrak{D}_{ij}(1)} \left[\frac{n_j}{m_i} D_i^T - \frac{n_i}{m_j} D_j^T \right] \quad (29\text{-}45)$$

This expression can be simplified considerably by making use of a special form of Eq. 29-40. The auxiliary conditions on the coefficients, $c_{i0}^{(h,k)}$ (Eq. 28-74) can be written in terms of the diffusion coefficients. The result is

$$\sum_i \{m_i m_h D_{ih}(1) - m_i m_k D_{ik}(1)\} = 0 \quad (29\text{-}46)$$

Making use of this, we find that Eq. 29-40 becomes[53]

$$\sum_{j \neq i} \frac{1}{\mathfrak{D}_{ij}(1)} \{n_i m_h D_{jh}(1) - n_i m_k D_{jk}(1) - n_j m_h D_{ih}(1) + n_j m_k D_{ik}(1)\}$$

$$= (\delta_{ih} - \delta_{ik})\rho \quad (29\text{-}47)$$

This set of equations is not linearly independent and hence could not be used without Eq. 29-46 to obtain the coefficients. Nevertheless they are valid relations. Because of Eq. 29-47 and the fact that $\sum_k \mathbf{d}_k = 0$, Eq. 29-45 becomes

$$\sum_j \frac{n_i n_j}{\mathfrak{D}_{ij}(1)} (\bar{\mathbf{V}}_i - \bar{\mathbf{V}}_j) = -n^2 \mathbf{d}_i - (\nabla \ln T) \sum_j \frac{1}{\mathfrak{D}_{ij}(1)} \left[\frac{n_j}{m_i} D_i^T - \frac{n_i}{m_j} D_j^T \right]$$

$$(29\text{-}48)$$

This is a set of $(\nu - 1)$ independent equations which are often directly applicable to hydrodynamic problems.

2. The Coefficient of Thermal Diffusion. To obtain the coefficients of thermal diffusion, it is necessary to evaluate the functions $\mathbf{A}_i$. In this case two terms in the Sonine polynomial expansion must be considered. Use of a single term results in a zero thermal diffusion coefficient; for this reason thermal diffusion is frequently referred to as a "second order" effect. In this case, Eq. 28-75 becomes

$$\sum_j \sum_{m'=0}^{1} \bar{Q}_{ij}^{mm'} a_{jm'}(2) = -R_{im} \quad (29\text{-}49)$$

In terms of the $\Omega^{(l)}(s)$

$$\bar{Q}_{ij}^{00} = 8 \sum_k \frac{n_k m_k}{\sqrt{m_i m_j}\,(m_i + m_k)} [n_i m_i(\delta_{ij} - \delta_{jk}) - n_j m_j(1 - \delta_{ik})] \Omega_{ik}^{(1)}(1)$$

$$(29\text{-}50)$$

[53] This set of equations is a special case of the general relations (Eq. 28-70).

$$\tilde{Q}^{01}_{ij} = -8\left(\frac{m_i}{m_j}\right)^{\frac{3}{2}} \sum_k \frac{n_i n_k m_k^2}{(m_i + m_k)^2} (\delta_{ij} - \delta_{kj}) \left[\Omega^{(1)}_{ik}(2) - \frac{5}{2}\,\Omega^{(1)}_{ik}(1)\right] \qquad (29\text{-}51)$$

$$\tilde{Q}^{10}_{ij} = \frac{m_j}{m_i}\,\tilde{Q}^{01}_{ij} \qquad (29\text{-}52)$$

$$\tilde{Q}^{11}_{ij} = 8\left(\frac{m_i}{m_j}\right)^{\frac{3}{2}} \sum_k \frac{n_i n_k m_k}{(m_i + m_k)^3} \left[(\delta_{ij} - \delta_{jk}) \left[\begin{array}{l} \frac{5}{4}(6m_j^2 + 5m_k^2)\,\Omega^{(1)}_{ik}(1) \\ - 5m_k^2 \Omega^{(1)}_{ik}(2) \\ + m_k^2 \Omega^{(1)}_{ik}(3) \end{array} \right] + (\delta_{ij} + \delta_{jk}) 2 m_j m_k \Omega^{(2)}_{ik}(2) \right] \qquad (29\text{-}53)$$

$$R_{im} = \delta_{m1} \frac{15}{4}\, n_i \sqrt{\frac{2kT}{m_i}} \qquad (29\text{-}54)$$

The expressions in Eq. 29-49 form a set of linear equations for the coefficients a_{j0} and a_{j1}, which can be solved by Cramer's rule.[54] Then from Eq. 29-9 one obtains the expression for the coefficient of thermal diffusion:

$$D_i^T(2) = n_i \sqrt{\frac{m_i kT}{2}} \; \frac{\begin{vmatrix} \tilde{Q}^{00}_{11} & \tilde{Q}^{00}_{12} & \cdots & \tilde{Q}^{00}_{1\nu} & \tilde{Q}^{01}_{11} & \tilde{Q}^{01}_{12} & \cdots & \tilde{Q}^{01}_{1\nu} & 0 \\ \tilde{Q}^{00}_{21} & \tilde{Q}^{00}_{22} & \cdots & \tilde{Q}^{00}_{2\nu} & \tilde{Q}^{01}_{21} & \tilde{Q}^{01}_{22} & \cdots & \tilde{Q}^{01}_{2\nu} & 0 \\ \cdot & \cdot & \cdots & \cdot & \cdot & \cdot & \cdots & \cdot & \cdot \\ \cdot & \cdot & \cdots & \cdot & \cdot & \cdot & \cdots & \cdot & \cdot \\ \cdot & \cdot & \cdots & \cdot & \cdot & \cdot & \cdots & \cdot & \cdot \\ \tilde{Q}^{00}_{\nu 1} & \tilde{Q}^{00}_{\nu 2} & \cdots & \tilde{Q}^{00}_{\nu\nu} & \tilde{Q}^{01}_{\nu 1} & \tilde{Q}^{01}_{\nu 2} & \cdots & \tilde{Q}^{01}_{\nu\nu} & 0 \\ \tilde{Q}^{10}_{11} & \tilde{Q}^{10}_{12} & \cdots & \tilde{Q}^{10}_{1\nu} & \tilde{Q}^{11}_{11} & \tilde{Q}^{11}_{12} & \cdots & \tilde{Q}^{11}_{1\nu} & R_{11} \\ \tilde{Q}^{10}_{21} & \tilde{Q}^{10}_{22} & \cdots & \tilde{Q}^{10}_{2\nu} & \tilde{Q}^{11}_{21} & \tilde{Q}^{11}_{22} & \cdots & \tilde{Q}^{11}_{2\nu} & R_{21} \\ \cdot & \cdot & \cdots & \cdot & \cdot & \cdot & \cdots & \cdot & \cdot \\ \cdot & \cdot & \cdots & \cdot & \cdot & \cdot & \cdots & \cdot & \cdot \\ \cdot & \cdot & \cdots & \cdot & \cdot & \cdot & \cdots & \cdot & \cdot \\ \tilde{Q}^{10}_{\nu 1} & \tilde{Q}^{10}_{\nu 2} & \cdots & \tilde{Q}^{10}_{\nu\nu} & \tilde{Q}^{11}_{\nu 1} & \tilde{Q}^{11}_{\nu 2} & \cdots & \tilde{Q}^{11}_{\nu\nu} & R_{\nu 1} \\ \delta_{i1} & \delta_{i2} & \cdots & \delta_{i\nu} & 0 & 0 & \cdots & 0 & 0 \end{vmatrix}}{\begin{vmatrix} \tilde{Q}^{00}_{11} & \cdots & \tilde{Q}^{00}_{1\nu} & \tilde{Q}^{01}_{11} & \cdots & \tilde{Q}^{01}_{1\nu} \\ \cdot & \cdots & \cdot & \cdot & \cdots & \cdot \\ \cdot & \cdots & \cdot & \cdot & \cdots & \cdot \\ \cdot & \cdots & \cdot & \cdot & \cdots & \cdot \\ \tilde{Q}^{00}_{\nu 1} & \cdots & \tilde{Q}^{00}_{\nu\nu} & \tilde{Q}^{01}_{\nu 1} & \cdots & \tilde{Q}^{01}_{\nu\nu} \\ \tilde{Q}^{10}_{11} & \cdots & \tilde{Q}^{10}_{1\nu} & \tilde{Q}^{11}_{11} & \cdots & \tilde{Q}^{11}_{1\nu} \\ \cdot & \cdots & \cdot & \cdot & \cdots & \cdot \\ \cdot & \cdots & \cdot & \cdot & \cdots & \cdot \\ \cdot & \cdots & \cdot & \cdot & \cdots & \cdot \\ \tilde{Q}^{10}_{\nu 1} & \cdots & \tilde{Q}^{10}_{\nu\nu} & \tilde{Q}^{11}_{\nu 1} & \cdots & \tilde{Q}^{11}_{\nu\nu} \end{vmatrix}} \qquad (29\text{-}55)$$

[54] Cramer's rule states that the solution of a set of linear equations can be expressed as the ratio of determinants of the coefficients in the usual manner.

3. The Coefficient of Viscosity. For a ν component mixture, the first approximation to the multicomponent viscosity is (according to Eq. 29-20)

$$\mu(1) = \tfrac{1}{2}kT \sum_j n_j b_{j0}(1) \tag{29-56}$$

The $b_{j0}(1)$ are then determined by the ν equations (according to Eq. 28-75)

$$\sum_j \left(\frac{Q_{ij}^{00}}{R_{i0}}\right) b_{j0}(1) = -1, \qquad i = 1, 2, 3, \cdots, \nu \tag{29-57}$$

with

$$Q_{ij}^{00} = \sum_l n_i n_l \{\delta_{ij}[\mathbf{W}_i; \mathbf{W}_i]_{il} + \delta_{jl}[\mathbf{W}_i; \mathbf{W}_l]_{il}\} \tag{29-58}$$

In this case $\mathbf{W}_i$ and R_{i0} are

$$\mathbf{W}_i = \mathbf{W}_i\mathbf{W}_i - \tfrac{1}{3}W_i^2\mathbf{1} \tag{29-59}$$

$$\begin{aligned} R_{i0} &= \int 2(\mathbf{W}_i : \mathbf{W}_i) f_i^{(0)} d\mathbf{V}_i \\ &= -\tfrac{4}{3} \int W_i^4 f_i^{(0)} d\mathbf{V}_i = -5n_i \end{aligned} \tag{29-60}$$

From Eq. 29-57 one can, of course, obtain the $b_{j0}(1)$ as ratios of two determinants of order ν by Cramer's rule. However, the only quantity which appears in the expression for the coefficient of viscosity is $\sum_j n_j b_{j0}(1)$. This can be written as a ratio of two determinants—the one in the numerator being of order $(\nu + 1)$, and that in the denominator of order ν. Specifically, one obtains,

$$[\mu]_1 = \frac{\begin{vmatrix} J_{11} & J_{12} & \cdots & J_{1\nu} & 1 \\ J_{21} & J_{22} & \cdots & J_{2\nu} & 1 \\ \cdot & \cdot & & \cdot & \\ \cdot & \cdot & & \cdot & \\ \cdot & \cdot & & \cdot & \\ J_{\nu 1} & J_{\nu 2} & \cdots & J_{\nu\nu} & 1 \\ 1 & 1 & \cdots & 1 & 0 \end{vmatrix}}{|J_{ij}|} \tag{29-61}$$

with

$$J_{ij} = \frac{2Q_{ij}^{00}}{n_j kT R_{i0}} = -\frac{2}{5}\frac{1}{kT}\sum_l \frac{n_l}{n_j}\{\delta_{ij}[\mathbf{W}_i; \mathbf{W}_i]_{il} + \delta_{jl}[\mathbf{W}_i; \mathbf{W}_l]_{il}\} \tag{29-62}$$

In terms of the $\Omega_{ij}^{(l)}(s)$

$$J_{ij} = -\frac{32}{15}\frac{m_i}{n_j m_j kT}\sum_l \frac{n_l m_l}{(m_i + m_l)^2}\left[\begin{array}{l} 5m_j(\delta_{ij} - \delta_{jl})\Omega_{il}^{(1)}(1) \\ + \tfrac{3}{2}m_l(\delta_{ij} + \delta_{jl})\Omega_{il}^{(2)}(2) \end{array}\right] \tag{29-63}$$

This result can easily be extended to include the effect of more terms in the Sonine polynomial expansion. The formal results can be simplified considerably in the case of binary mixtures and pure gases. (See Sec. D.)

4. The Coefficient of Thermal Conductivity. The expression for the energy flux is usually written in terms of the diffusion velocities and the temperature gradient. Combining Eq. 29-30 and 29-48 gives

$$\mathbf{q} = -\lambda\nabla T + \frac{5}{2}kT\sum_i n_i\overline{\mathbf{V}}_i + \frac{kT}{n}\sum_{i,j}\frac{n_j D_i^T}{m_i\mathfrak{D}_{ij}}(\overline{\mathbf{V}}_i - \overline{\mathbf{V}}_j) \qquad (29\text{-}64)$$

In this expression

$$\lambda = \lambda' - \frac{k}{2n}\sum_{i,j}\frac{n_i n_j}{\mathfrak{D}_{ij}}\left[\frac{D_i^T}{n_i m_i} - \frac{D_j^T}{n_j m_j}\right]^2 \qquad (29\text{-}65)$$

is the usual coefficient of thermal conductivity. The quantity λ' is expressed in terms of the Sonine expansion coefficients by Eq. 29-33 from which, on applying the methods and results described above, we obtain

$$\lambda'(2) = -\tfrac{75}{8}k^2T\,\frac{\begin{vmatrix} q_{11}^{00} & \cdots & q_{1\nu}^{00} & q_{11}^{01} & \cdots & q_{1\nu}^{01} & 0 \\ \vdots & & \vdots & \vdots & & \vdots & \vdots \\ q_{\nu 1}^{00} & \cdots & q_{\nu\nu}^{00} & q_{\nu 1}^{01} & \cdots & q_{\nu\nu}^{01} & 0 \\ q_{11}^{10} & \cdots & q_{1\nu}^{10} & q_{11}^{11} & \cdots & q_{1\nu}^{11} & 1 \\ \vdots & & \vdots & \vdots & & \vdots & \vdots \\ q_{\nu 1}^{10} & \cdots & q_{\nu\nu}^{10} & q_{\nu 1}^{11} & \cdots & q_{\nu\nu}^{11} & 1 \\ 0 & \cdots & 0 & 1 & \cdots & 1 & 0 \end{vmatrix}}{\begin{vmatrix} q_{11}^{00} & \cdots & q_{1\nu}^{00} & q_{11}^{01} & \cdots & q_{1\nu}^{01} \\ \vdots & & \vdots & \vdots & & \vdots \\ q_{\nu 1}^{00} & \cdots & q_{\nu\nu}^{00} & q_{\nu 1}^{01} & \cdots & q_{\nu\nu}^{01} \\ q_{11}^{10} & \cdots & q_{1\nu}^{10} & q_{11}^{11} & \cdots & q_{1\nu}^{11} \\ \vdots & & \vdots & \vdots & & \vdots \\ q_{\nu 1}^{10} & \cdots & q_{\nu\nu}^{10} & q_{\nu 1}^{11} & \cdots & q_{\nu\nu}^{11} \end{vmatrix}} \qquad (29\text{-}66)$$

Here

$$q_{ij}^{mm'} = \frac{\sqrt{m_i m_j}}{n_i n_j} \tilde{Q}_{ij}^{mm'} \tag{29-67}$$

where the $\tilde{Q}_{ij}^{mm'}$ are those given by Eq. 29-50, 51, 52, and 53.

B,30. Cited References and Bibliography.

Cited References

1. Born, M. *Atomic Physics*, 4th ed. Blackie, 1947.
2. Darwin, C. G. *The New Conceptions of Matter*. Macmillan, 1931.
3. Slater, J. C., and Frank, N. H. *Introduction to Theoretical Physics*. McGraw-Hill, 1933.
4. Uhlenbeck, G. E., and Goudsmit, S. *Naturwiss. 13*, 953 (1925); *Nature 117*, 264 (1926).
5. Mott, N. F., and Sneddon, I. N. *Wave Mechanics and Its Applications*. Oxford Univ. Press, 1948.
6. Pauling, L. *The Nature of the Chemical Bond*. Cornell, 1948.
7. Craig, D. P. *Proc. Roy. Soc. London A200*, 272 (1950a); *A200*, 390 (1950b); *A200*, 401 (1950c); *A200*, 474 (1950d).
8. Coulson, C. A. *Quart. Revs. London 1*, 144 (1947). See also the bibliography in *Proc. Roy. Soc. London A207*, 1951.
9. Mulliken, R. S. *J. chim. phys. 46*, 497 (1949). Annual Reports of the Spectroscopic Laboratory, University of Chicago.
10. Lennard-Jones, J. *Proc. Roy. Soc. London A198*, 14 (1949). *Trans. Faraday Soc. 25*, 668 (1929).
11. Sherman, A., and Van Vleck, J. H. *Revs. Mod. Phys. 7*, 167–228 (1935).
12. Mulligan, J. *J. Chem. Phys. 19*, 347–362 (1951).
13. Roothaan, C. C. J. *Ph.D. Thesis*, Univ. of Chicago, 1950; published in *ONR Technical Report from the Univ. of Chicago* for the period Sept. 1, 1948 to May 31, 1949, Part Two; Report for the period June 1, 1949 to March 31, 1950, Part Two.
14. Herzberg, G. *Infrared and Raman Spectra*. Van Nostrand, 1945.
15. Gordy, W. *Revs. Mod. Phys. 20*, 668–717 (1948).
16. Bichowsky, F. R., and Rossini, F. D. *Thermochemistry of the Chemical Substances*. Reinhold, 1936. Also: Rossini, F. D., Wagman, D. D., Evans, W. H., Levine, S., and Jaffe, I. *Selected Values of Chemical Thermodynamic Properties*. Natl. Bur. Standards Circ. 500 U. S. Government Printing Office, 1952.
17. Herzberg, G. *Spectra of Diatomic Molecules*, 2nd ed. Van Nostrand, 1950.
18. Gaydon, A. G. *Dissociation Energies*. Dover, 1950.
19. Szwarc, M., and Evans, M. G. *J. Chem. Phys. 18*, 618–622 (1950).
20. Szwarc, M. *Chem. Revs. 47*, 75 (1950).
21. Kistiakowsky, G. B., and Van Artsdalen, E. R. *J. Chem. Phys. 12*, 469 (1944).
22. Prosen, E., Johnson, W., and Rossini, F. *J. Research Natl. Bur. Standards 37*, 51 (1946).
23. Steacie, E. W. R. *Atomic and Free Radical Reactions*. Reinhold, 1946.
24. Syrkin, Y. K., and Dyatkina, M. E. *Structure of Molecules*. Interscience, 1950.
25. A discussion on bond energies and bond lengths. *Proc. Roy. Soc. London A207* (*A1088*), 1–136 (1951).
26. Tolman, R. C. *Statistical Mechanics with Applications to Chemistry and Physics*. Chemical Catalog, 1927.

27a. Hinshelwood, C. N. *The Kinetics of Chemical Change*. Oxford Univ. Press, 1940.

27b. Laidler, K. J. *Chemical Kinetics*. McGraw-Hill, 1950.

28. Eyring, H., and Polanyi, M. *Z. physik. Chem. B12*, 279 (1931).
29. Wigner, E. *Trans. Faraday Soc. 34*, 29 (1938).
30. Evans, M. G., and Polanyi, M. *Trans. Faraday Soc. 34*, 11 (1938).
31. Glasstone, S., Laidler, K. J., and Eyring, H. *Theory of Rate Processes*. McGraw-Hill, 1941.

32. de Boer, J. *Repts. Progr. in Phys. 12*, 305 (1949).
33. Tolman, R. H. *Principles of Statistical Mechanics*. Oxford Univ. Press, 1938.
34. Hirschfelder, J. O., Curtiss, C. F., and Bird, R. B. *The Molecular Theory of Gases and Liquids*. Wiley, 1954.
35. Fowler, R. H., and Guggenheim, E. A. *Statistical Thermodynamics*. Cambridge Univ. Press, 1939.
36. Mayer, J. E., and Mayer, M. G. *Statistical Mechanics*. Wiley, 1940.
37. Grad, H. *Commun. on Pure and Appl. Math. 2*, 331 (1949).
38. Chapman, S., and Cowling, T. G. *The Mathematical Theory of Non-Uniform Gases*. Cambridge Univ. Press, 1939, 1952.
39. Kirkwood, J. G. *J. Chem. Phys. 15*, 72 (1947).
40. Kirkwood, J. G. *J. Chem. Phys. 18*, 817 (1950).
41. Enskog, D. *Arkiv Mat., Astron. Fysik Stockholm 16*, 16 (1922); *Kinetische Theorie der Vorgänge in mässig verdünnten Gasen. (Inaugural Disseration.)* Uppsala, Sweden. Almquist and Wiksell, 1917.
42. Curtiss, C. F., and Hirschfelder, J. O. *J. Chem. Phys. 17*, 550 (1949).

Bibliography

Eyring, H., Walter, J., and Kimball, G. E. *Quantum Chemistry*. Wiley, 1944.
Pauling, L. C., and Wilson, E. B., Jr. *Introduction to Quantum Mechanics*, 1st ed. McGraw-Hill, 1935.
Rice, F. O., and Teller, E. *The Structure of Matter*. Wiley, 1949.
Rojanski, V. B. *Introductory Quantum Mechanics*. Prentice-Hall, 1938.
Schiff, L. I. *Quantum Mechanics*. McGraw-Hill, 1949.
Tolman, R. C. *The Principles of Statistical Mechanics*. Oxford, 1938.

Lightning Source UK Ltd.
Milton Keynes UK
UKHW020409110122
396919UK00003B/201

9 780691 625669